COSMIC COLLISIONS

THE HUBBLE ATLAS OF MERGING GALAXIES

LARS LINDBERG CHRISTENSEN, RAQUEL YUMI SHIDA & DAVIDE DE MARTIN

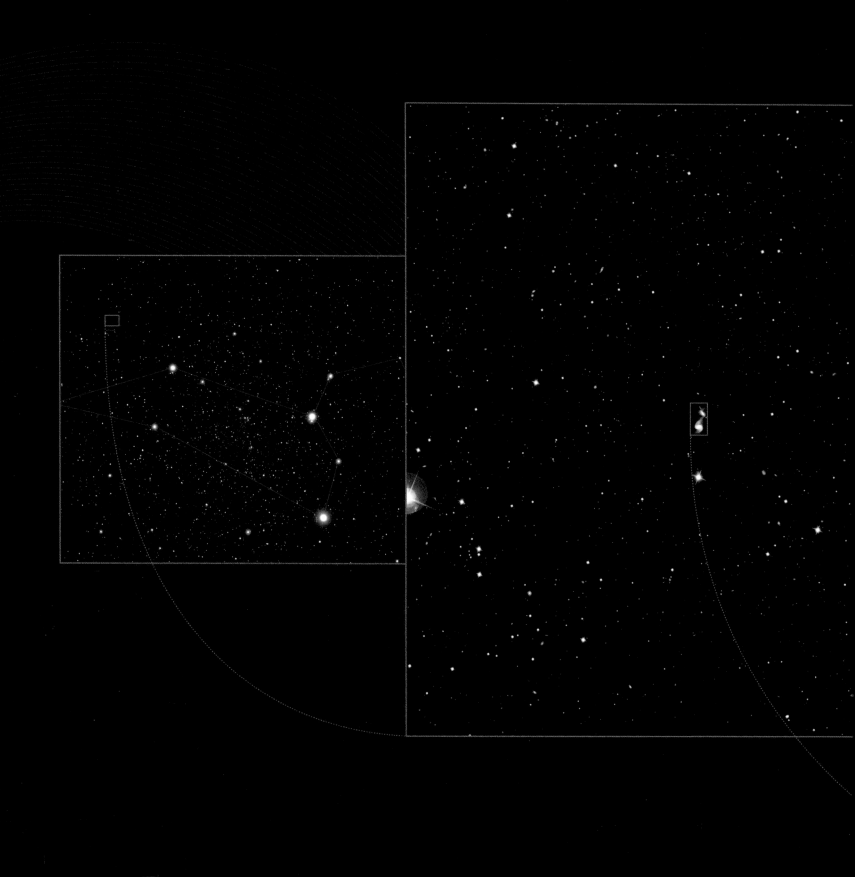

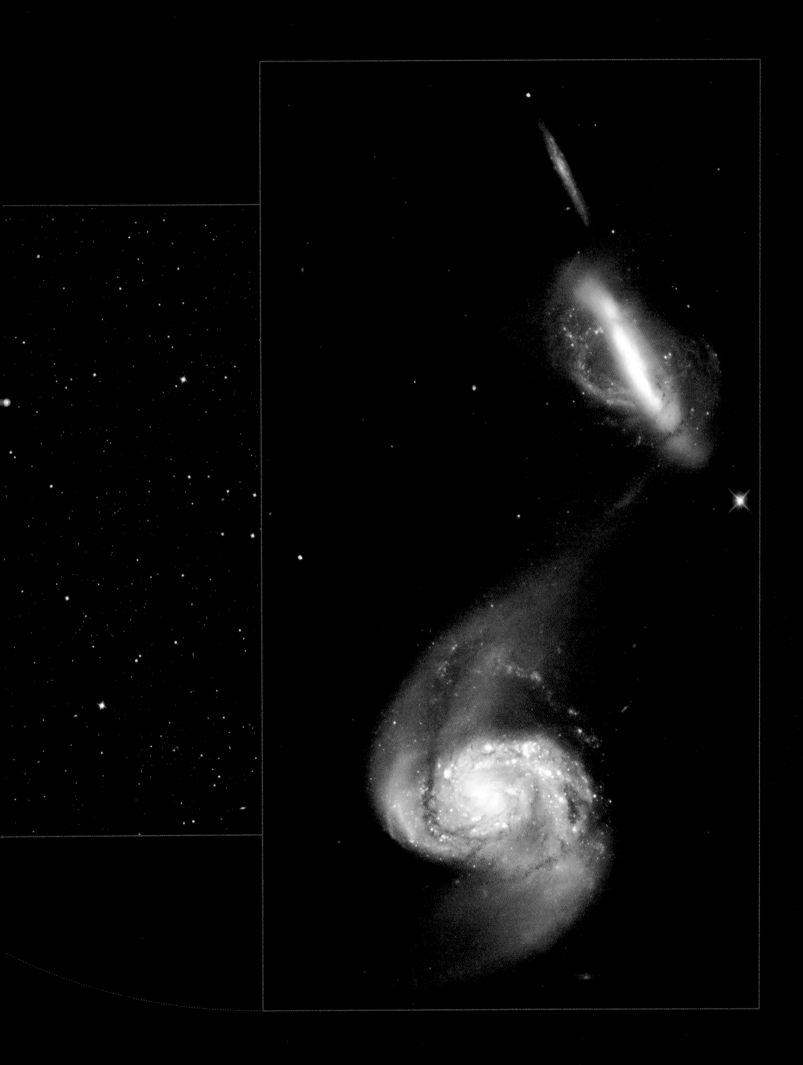

COSMIC COLLISIONS
The Hubble Atlas of Merging Galaxies

The Authors
Lars Lindberg Christensen
Raquel Yumi Shida
ESO
Garching, Germany

Davide De Martin
skyfactory.org
Venice, Italy

Design and Layout
Raquel Yumi Shida &
Martin Kornmesser
ESO
Garching, Germany

All books published by Springer are carefully produced. Nevertheless, authors, editors, and publisher do not warrant the information contained in these books, including this book, to be free of errors. Readers are advised to keep in mind that statements, data, illustrations, procedural details or other items may inadvertently be inaccurate.

ISBN: 978-0-387-93853-0
e-ISBN: 978-0-387-93855-4

Library of Congress Control Number: 2009930652

Springer Science + Business Media, LLC, © 2009

All rights reserved. This work may not be translated or copied in whole or in part without the written permission of the publisher (Springer Science+Business Media, LLC, 233 Spring Street, New York, NY 10013, USA) except for brief excerpts in connection with reviews or scholarly analysis. Use in connection with any form of information storage and retrieval, electronic adaptation, computer software, or by similar or dissimilar methodology now known or hereafter developed is forbidden. The use in this publication of trade names, trademarks, service marks, and similar terms, even if they are not identified as such, is not to be taken as an expression of opinion as to whether or not they are subject to proprietary rights.

Printed on acid-free paper.
9 8 7 6 5 4 3 2 1
springer.com

ZOOMING IN ON ARP 87 (INSIDE FRONT COVER)

To give an impression of the scale of the many Hubble images in this book here we show a "zoom" on Arp 87. The first image (left) shows the full constellation of Leo as taken with a regular camera. The second image (middle) shows a field of view from the Digitized Sky Survey centered on Arp 87. The third image (right) shows a Hubble image of Arp 87, roughly 2.3 arc-minutes across (1/25 degree). It is evident that Hubble's view is very narrow and that only small areas of the sky can be imaged at any one time. Arp 87 is a stunning pair of interacting galaxies. Stars, gas, and dust flow from the large spiral galaxy, NGC 3808 (the larger of the two galaxies) wrapping an enveloping arm around its companion NGC 3808A above. NGC 3808 is a nearly face-on spiral galaxy with a bright ring of star formation and several prominent dust arms. NGC 3808A is a spiral galaxy seen edge-on and is surrounded by a rotating ring that contains stars and interstellar gas clouds. The ring is perpendicular to the plane of the host galaxy disk and is called a "polar ring." The shapes of both galaxies have been distorted by their gravitational interaction. Arp 87 is approximately 300 million light-years away from Earth. As is also seen in other similar interacting galaxies, the corkscrew shape of the tidal material suggests that stars and gas drawn from the larger galaxy have been caught in the gravitational pull of the smaller one.

THE TADPOLE GALAXY (INSIDE BACK COVER)

Against a stunning backdrop of thousands of galaxies, this odd-looking galaxy with an enormous tidal streamer of stars appears to be racing through space, like a runaway pinwheel firework. Dubbed the "Tadpole," this spiral galaxy is unlike the textbook images of stately galaxies. Its distorted shape was caused by a small interloper, a very blue, compact galaxy visible in the upper left corner of the more massive Tadpole. The Tadpole resides about 420 million light-years away in the constellation of Draco.

Lars Lindberg Christensen, Raquel Yumi Shida & Davide De Martin

ARP 299

This system consists of a pair of galaxies, dubbed IC 694 and NGC 3690, which made a close pass some 700 million years ago. As a result of this interaction, the system underwent a fierce burst of star formation. In the last fifteen years or so six supernovae have popped off in the outer reaches of the galaxy, making this system a distinguished supernova factory. Arp 299 belongs to the family of ultraluminous infrared galaxies and is located in the constellation of Ursa Major, the Great Bear, approximately 150 million light-years away. Despite the enormous amount of absorbing dust its vigorous star formation makes it shine brightly in the ultraviolet.

TABLE OF CONTENTS

	Preface	7
1	Galaxies: The Big Picture	9
2	How Do Galaxies Form and Evolve?	41
3	Galaxy Collisions	53
4	The Colliding Galaxies Movie	71
5	The End	91
6	Gallery	99
	The Authors	136
	Resources	138
	Image Credits	139

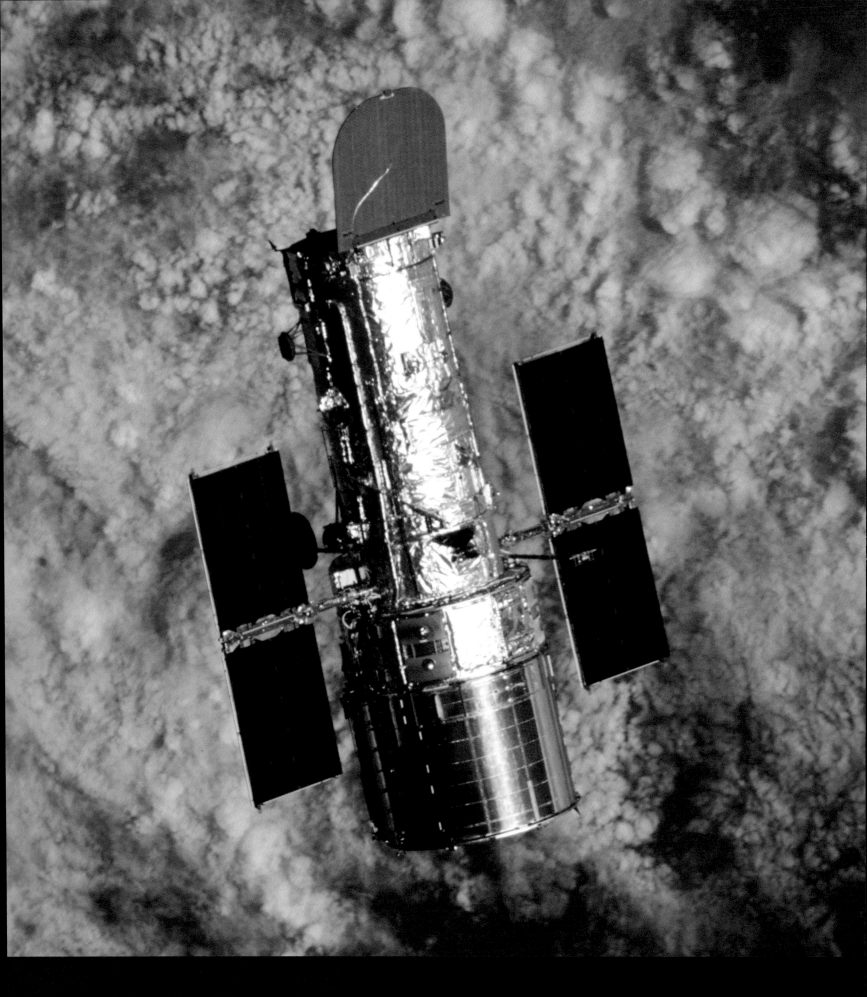

THE HUBBLE SPACE TELESCOPE

The Hubble Space Telescope has orbited our planet more than 100,000 times since 1990 and is still delivering new and fascinating astronomical images that answer some of the most fundamental questions about our origin.

PREFACE

The Hubble Space Telescope, a joint project of the European Space Agency (ESA) and NASA, has made some of the most important discoveries in the history of science. From its vantage point 600 km above Earth's surface, freed of the distorting effects of Earth's atmosphere, Hubble's "eyes" can see five times more sharply than those of ground-based telescopes and look deep into space to probe the profound mysteries that are still buried in the mists of time.

Among the most fascinating and dramatic events that Hubble has been able to show in high resolution are the cosmic collisions between galaxies. These gigantic encounters create phenomena that give rise to strange features involving clusters bursting with new stars, twisting lanes of gas and dust, and tidal tails extending over hundreds of thousands of light-years.

The importance of these cosmic encounters reaches far beyond the aesthetic Hubble images we present here. These collisions may be some of the most important processes shaping the universe we inhabit today. Colliding galaxies very likely hold some of the most important clues to our galactic ancestry and to our destiny. It now seems clear that the galaxy we all live in, the Milky Way, is still continuously undergoing merging events, both major and minor, and that this process is much more important in the lives of galaxies than previously thought.

Hubble's images are snapshots in time and have caught the individual stages of intergalactic collisions. These different stages can be put together into a movie (Chapter 4) showing how these monumental collisions progress.

In this book, we will give a brief and up-to-date introduction to the lives of galaxies — how they were born, evolve over time, and collide — using the best pictures taken by the Hubble Space Telescope. Many of these images are from a huge investigation of luminous infrared galaxies called the GOALS project (Great Observatory All-sky LIRG Survey, goals.ipac.caltech.edu). The Hubble observations were led by Aaron S. Evans from Stony Brook University (U.S.).

We would like to thank Bill Keel for providing inspiration and advice, and we are grateful to him for sharing his deep knowledge of the field. Aaron S. Evans and the GOALS team are to be thanked for taking the amazing Hubble data that made these vistas possible, and Maury Solomon and Harry Blom from Springer for taking on this project. Thanks also to Colleen Sharkey, Henrik Spoon and Bob Fosbury for valuable discussions and suggestions to the text.

Lars Lindberg Christensen, Raquel Yumi Shida & Davide De Martin

Munich and Venice, 1 March 2009

1 GALAXIES: THE BIG PICTURE

NGC 5866

This is a unique Hubble view of the disk galaxy NGC 5866 tilted nearly edge-on to our line of sight. Hubble's sharp vision reveals a crisp dust lane dividing the galaxy into two halves. The image highlights the galaxy's structure: a subtle, reddish bulge surrounding a bright nucleus, a blue disk of stars running parallel to the dust lane, and a transparent outer halo. Some faint, wispy trails of dust can be seen meandering away from the disk of the galaxy out into the bulge and inner halo of the galaxy. The outer halo is dotted with numerous globular star clusters, with nearly a million stars in each. Background galaxies that are millions to billions of light-years farther away are also seen through the halo.

Essentially everything we can see in the night sky — whether with the unaided eye or with telescopes — are galaxies, or parts of galaxies. When thinking about galaxies, we need to think BIG: a galaxy is an almost unfathomably large collection of dust, gas, dark matter, planets, and stars — billions of them, packed together by the force of gravity.

> *"Mapping the Milky Way is much like trying to map a crowded, foggy city from a single vantage point inside the murk."*

This book is about interacting galaxies, but what is a galaxy? The galaxy we know best is, not surprisingly, the one we live in — the Milky Way. From our vantage point inside our galaxy, it is difficult to have a clear idea of how the Milky Way might look from the outside. Mapping the Milky Way is much like trying to map a crowded, foggy city from a single vantage point inside the murk. It took humankind a long time to establish the existence of the Milky Way and to recognize and acknowledge the existence of other galaxies in the universe.

Discovery of the Milky Way

If you have ever had the opportunity to observe the night sky from a place without light pollution, you will certainly have noticed regions where the concentration of stars is higher — looking like a chalky band crossing the sky behind the stars. When seen from a really dark site, the Milky Way is one of the most amazing displays of nature.

In ancient times there was no scientific explanation for this chalky band. The Vikings saw it as a road walked by the dead to reach heaven; for the Incas it was a river from which the weather god Apu Illapu drew water to make rain; the Egyptians thought it was a great river in the sky or "the Nile in the Heaven." The Milky Way takes its name from the Latin *Via Lactea*, in turn derived from the Greek Γαλαξιας (Galaxias). In Greek mythology, Hermes, the messenger of the gods, attempted to make the infant Hercules immortal by letting him suckle at the breast of the sleeping goddess Hera, but Hera woke and thrust the baby away, tearing her breast from the baby's mouth, so that her milk spurted across the sky to form the Milky Way.

Early scientists could not place the Milky Way in their world view, so they tended to ignore it. The first to get close to the true nature of the Milky Way was the Italian Galileo Galilei, who observed it through his newly invented telescope in 1609 and 1610. He was

THE ORIGIN OF THE MILKY WAY

According to Greek mythology the Milky Way was created when the infant Hercules tried to suckle from Hera's breast. The story is interpreted here by the Italian painter Jacopo Comin, better known as Tintoretto. The original painting is exhibited at the National Gallery, London.

probably the first person to notice that the patches in the Milky Way were formed from many stars that are apparently very close to each other, noting in *Sidereus Nuncius* that the Milky Way was *"nothing other than a mass of innumerable stars scattered in clusters."* However, not even Galileo fully understood the implications of his discovery, and he did not try to explain further the strange and uneven distribution of the stars across the sky.

The first person to try to find an explanation was the English amateur astronomer Thomas Wright. In 1750 Wright published a book in which he explained the appearance of the Milky Way as *"an optical effect due to our immersion in what locally approximates to a flat layer of stars."*

The great philosopher Immanuel Kant took up Wright's idea and elaborated on it. In his book, the *Universal Natural History and Theory of Heaven*, written in 1755, he arrived at a surprisingly accurate picture. Kant believed that the Milky Way was a large flat disk of stars with a central bulge. Kant also thought that this huge system had to be just

THE MILKY WAY

This is how a night-time observer on Earth sees the Milky Way, from the inside. If we could see the Milky Way from the outside it would look roughly like any other spiral galaxy.

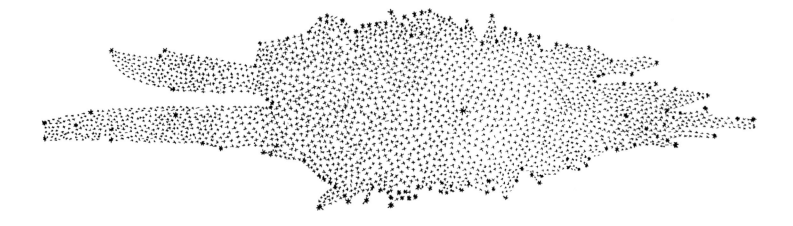

HERSCHEL'S MILKY WAY

A great pioneer in the study of the stars, William Herschel was appointed private astronomer to the King of England in recognition of his discovery of the planet Uranus in 1781. In 1785 he made his first, not entirely successful, attempt at a scientific explanation of the structure of the Milky Way Galaxy from star counts. A big step forward, although he incorrectly deduced that the Solar System was near the center of the Milky Way.

one of countless similar systems, separated by immense tracts of space. He called these systems "island universes." And this is what galaxies are: islands floating in the large cosmic ocean, each filled with myriads of stars. Without making any scientific observations, Kant had started an epoch of profound change in astronomical thinking.

The first scientific observations were made by the eminent English astronomer William Herschel. Herschel studied how the stars were scattered across the sky by counting them carefully in certain regions. As a result, he came to the conclusion that the Milky Way was in the shape of a disk. The stars counted by Herschel were limited to those he could observe with his telescope and as such, only account for part of the story.

In the two hundred years that followed, the collective efforts of astronomers and the progress of telescope technology allowed humankind to peer farther and farther out beyond the Milky Way to reveal that the milky band in the night sky is merely the local view you get from within a large galaxy.

Measuring and Weighing the Milky Way

The Milky Way is a reasonably luminous spiral system and contains two to four hundred billion stars, many like our own Sun. We can only see the nearest or brightest 6,000 of them with our unaided eyes because we cannot see through the clouds of dust and gas that pervade the Milky Way and block the light from more distant stars. Although two hundred billion is an almost inconceivable number, it is only the beginning. Astronomers believe there are more than a hundred billion galaxies in the universe.

The mass of the stars in the Milky Way is thought to be about 400 billion solar masses. However, even more mass is probably hidden in the form of dark matter (see box on page 47), spread relatively evenly into a dark matter halo with at least an additional

> *" Galaxies are islands floating in the large cosmic ocean, each filled with myriads of stars. "*

600 billion solar masses. There is a considerable uncertainty in this estimate, which stems from the difficulty astronomers have in measuring the mass of a mystery object of unknown extent and content at a distance of hundreds of thousands of light-years.

The visible part of the Milky Way galaxy measures about 100,000 light-years across and is about 1,000 light-years thick. This is — more or less — in the same proportions as an old-fashioned LP vinyl disk. The Milky Way has a central bulge some 10,000 to 12,000 light-years thick, and the whole huge body is rotating.

Our Sun, along with its small system of planets, moons, dwarf planets, asteroids, and comets, lies in the so called Orion Arm, some 26,000 light-years from the center.

To gain some idea about the sheer size of the Milky Way, imagine a scale model where 1 cm is equal to 10 million km. On this scale, the Sun would be the size of a grain of rice, and Earth would be a microscopic speck of dust about 15 cm from the Sun. The whole Solar System would be just 12 meters in diameter, and the nearest star, Proxima Centauri, would be another rice grain at a distance of 40 km. But on this scale, the Milky Way's disk would have a diameter of a million km. A speck of dust near a rice grain in a disk one and a half times greater than the Moon's orbit — that's our place in the Milky Way!

One of the most surprising facts about galaxies is how empty they are. We have seen that star systems scale to the size of rice grains, separated by tens of empty kilometers, but let us make another imaginary experiment, scaling so that the whole Solar System is now the size of a small grain of rice. Imagine collecting all the matter in the Milky Way in a single sphere with the same density as the Sun (without worrying about how it could be done, let alone who would do it — this is the joy of thought experiments.) Then this sphere, made up of roughly 1,000 billion Suns, would have a diameter of only a little more than twice the distance of the Sun to Pluto, or 15 billion km. The diameter of the Milky Way (now empty) would correspond to the distance between Milan and Turin in Italy (140 km, or just over an hour by car). Galaxies are very empty indeed.

> *" A speck of dust near a rice grain in a disk one and a half times greater than the Moon's orbit — that's our place in the Milky Way! "*

MILKY WAY MAP

This illustration represents the structure of our Milky Way as it is known today. The Milky Way's different spiral arms and its central bar are seen (in yellow). The location of the Sun is marked as a green dot.

THE PINWHEEL GALAXY

The gigantic Pinwheel Galaxy (Messier 101), one of the best known examples of "grand design spirals," and its supergiant star-forming regions. The giant spiral disk of stars, dust, and gas is 170,000 light-years across, or nearly twice the diameter of our Milky Way. The galaxy is estimated to contain at least one trillion stars. Approximately 100 billion of these stars alone might be like our Sun in terms of temperature and lifetime. Hubble's high resolution reveals millions of the galaxy's individual stars in this image.

> *"The Great Debate was arguably as important to our world view as some of the discoveries made by the explorers at sea in the seventeenth century."*

Other Island Universes

Until the twentieth century, the Milky Way was considered to be the whole universe. Observations did show some strange spiral nebulae scattered in the sky, such as the Triangulum Galaxy, the Pinwheel Galaxy (page 17), and others, but astronomers considered that these spiral nebulae were just simple gaseous objects inside the Milky Way.

Early in the twentieth century, things began to change, and some astronomers became convinced that Kant was right and the spiral nebulae were separate island universes, other Milky Ways. The controversy led to an historical debate on "The Scale of the Universe" among astronomers. The "Great Debate," as it became known, took place on April 26, 1920, in the Baird auditorium of the Smithsonian Museum of Natural History in New York. The Great Debate was arguably as important to our world view as, for instance, some of the discoveries made by the explorers at sea in the seventeenth century — a clear example of humanity struggling to find its place within the cosmic order.

The lines were drawn and the combatants in place, each armed with evidence that, due to the status of observations at the time, at best would be fragmentary and at worst plain wrong. On one side, Harlow Shapley, one of the most notable astronomers of his time, argued in favor of the Milky Way being the entirety of the universe. He correctly put the Solar System outside the center of the Milky Way, but believed that systems such as Andromeda and the other spiral nebulae were just nearby gas clouds and simply part of the Milky Way. He claimed that if Andromeda and the other nebulae were not part of the Milky Way, then the distance to them would be as much as hundreds of millions of light-years — a figure that most astronomers of the time found preposterous.

Heber D. Curtis, another prominent astronomer, for the other side, claimed that Andromeda and other such nebulae were separate galaxies, or "island universes." He supported his arguments with observations showing that there were more novae in Andromeda than in the Milky Way. Why would there be more novae in one small section of the galaxy than the others? This supported the idea of Andromeda as a separate galaxy,

> *"The protagonists left the auditorium in New York without a resolution to the immensely important problem of the scale of the universe."*

different in age from the Milky Way and with its own rate of occurrence of novae. Curtis presented other evidence, such as the dark lanes found in other galaxies, and argued that they were similar to the dust clouds found in our own Milky Way.

The protagonists left the auditorium in New York without a resolution to the immensely important problem of the scale of the universe, but the debate continued and strengthened when, in May 1921, the Bulletin of the National Research Council published two papers by Shapley and Curtis laying out their respective positions.

The first clue that Curtis was on the right track came from Vesto Melvin Slipher. As early as 1912, while working at the Lowell Observatory in Flagstaff, Arizona, Slipher had obtained spectra of some spiral nebulae and noted that the spectral lines of galaxies were shifted toward the red edge of the spectrum, demonstrating that these galaxies were receding from our point of view. Slipher also noted that the spiral nebulae had no emission lines (bright lines in their spectrum), as would have been expected if they were really the gaseous nebulae of the then-current theory.

Today, however, the astronomer Edwin Powell Hubble is the main person credited with resolving the issue. In 1919, George Ellery Hale offered Hubble a staff position at the Mount Wilson Observatory, where he continued to work until his death. Using the new 100-inch Hooker reflector telescope, the largest and most powerful instrument of the time, Hubble captured images of some of the spiral nebulae, resolving the single stars in them and identifying Cepheid variables in several objects, including the Andromeda Galaxy.

Cepheids are a particular type of star, discovered to be variable by John Goodricke in 1784. A Cepheid star varies its luminosity in a very regular way, expanding and contracting in pulses, like the beat of a giant cosmic heart. Cepheids have a very strong relation between the period of their variability and their real, or internal, brightness (known as

> *"The Milky Way is just one of many galaxies scattered throughout the universe."*

the absolute luminosity). A Cepheid with a three-day period has an absolute luminosity of about 800 times that of the Sun. A Cepheid with a thirty-day period is 10,000 times as bright as the Sun.

Because of this relationship (discovered and stated by astronomer Henrietta Swan Leavitt in 1908), and the high luminosity that makes Cepheids visible at great distances, a Cepheid variable can be used as a unique "standard candle" to determine the distance to its host galaxy.

Edwin Hubble, together with Milton L. Humason, used this relationship to measure the distances to some spiral nebulae, proving that they are millions of light-years distant — much farther than previously thought — and so confirming that they were actually galaxies, with the implication that the Milky Way is just one of many galaxies scattered throughout the universe. Hubble announced the discovery on January 1, 1925, adding more fuel to the fire of the ongoing debate and influencing the outcome decisively while fundamentally changing our view of the universe.

Different Types of Galaxies

After making the key observations that established the size of the universe and the true role of our Milky Way within it, Edwin Hubble tried to classify galaxies according to their shape in an attempt to discover patterns with physical meanings. In his book, published in 1936 and entitled *The Realm of the Nebulae* (the word "galaxy" only slowly became common during the 1930s), Hubble established a classification scheme for galaxies that today is called the "Hubble sequence," or the "Hubble tuning fork" because of its particular shape.

Hubble's classification system has proved surprisingly robust. Although this scheme is now considered somewhat too simple, the basic ideas still hold. It was thought only to describe the visual appearance of galaxies, but it is also well correlated with physically

NGC 4013 SEEN EDGE ON

This picture reveals with exquisite detail huge clouds of dust and gas extending along, as well as far above, NGC 4013's main disk. NGC 4013 is a spiral galaxy, similar to our Milky Way, lying some 55 million light-years from Earth in the direction of the constellation Ursa Major. Viewed face-on, it would look like a nearly circular pinwheel, but NGC 4013 happens to be seen edge-on from our vantage point.

> *"These oddball galaxies were largely ignored by astronomers until the mid-1950s, but have since turned out to be crucial for our understanding of how galaxies evolve."*

interesting parameters such as stellar content, gas content, and environment. The system was later extended by the astronomers Gérard de Vaucouleurs and Sidney van den Bergh and has been the most commonly used system for the classification of galaxies.

The diagram shows the shapes (or "morphology") of galaxies and is roughly divided into two parts: elliptical galaxies and spiral galaxies. Hubble gave the ellipticals numbers from zero to seven that characterize the apparent shape of the galaxy — "E0" is almost round, while "E7" is very elliptical. Elliptical galaxies show little to no structure and are generally characterized by old stellar populations and very little of the gas and dust needed to form new stars.

Spirals are subdivided into a sequence jointly defined by the degree of tightness of the spiral arms and the importance of the central bulge. "Sa" galaxies have a bright bulge and tightly wound arms, while "Sc" galaxies have loosely wound arms and a relatively less important core region. The central bulge generally contains older stars, similar to those seen in elliptical galaxies. There is a thin outer disk enclosing the bulge, with bright arms forming the characteristic spiral structure. Each of the spiral types exists in a version with and without a central bar. The barred spirals have a straight bar of stars that runs through the central bulge. The spiral arms in barred spirals usually start at the end of the bar instead of at the bulge. Barred spirals have a "B" in their classification. An "SBc" is thus a loosely wound barred spiral galaxy.

Hubble found that some galaxies were difficult to fit into his neatly arranged tuning-fork diagram. These include irregular galaxies with odd shapes, the very small dwarf galaxies, and the giant elliptical galaxies that are found at the centers of some clusters of galaxies. These oddball galaxies were largely ignored by astronomers until the mid-1950s, but have since turned out to be crucial for our understanding of how galaxies evolve.

For a time, the Hubble tuning fork was thought to be an evolutionary sequence such that galaxies might evolve from one type to another, progressing from left to right across the tuning-fork diagram. Hence elliptical galaxies, including S0, were called "early-type,"

M74, A GRAND-DESIGN SPIRAL

Messier 74, also called NGC 628, is a stunning example of a "grand-design" spiral galaxy that is viewed by Earth observers nearly face-on. Its nearly perfectly symmetrical spiral arms emanate from the central nucleus and are dotted with clusters of young blue stars. One can also see a smattering of bright pink regions decorating the spiral arms. These are huge, relatively short-lived, clouds of hydrogen gas that glow due to the strong radiation from hot, young stars embedded within them. These regions of star formation emit copious amounts of light at ultraviolet wavelengths and astronomers call them HII regions.

CLASSIFICATION OF GALAXIES

Edwin Hubble's galaxy classification system from 1936 is still in use today. For a time the Hubble tuning fork was thought to be an evolutionary sequence — that galaxies might evolve from one type to another, progressing from ellipticals (left) to spirals (right) across the tuning-fork diagram. In fact, current evidence suggests the opposite, with an important part of the story being that ellipticals form as a result of collisions and mergers between spirals.

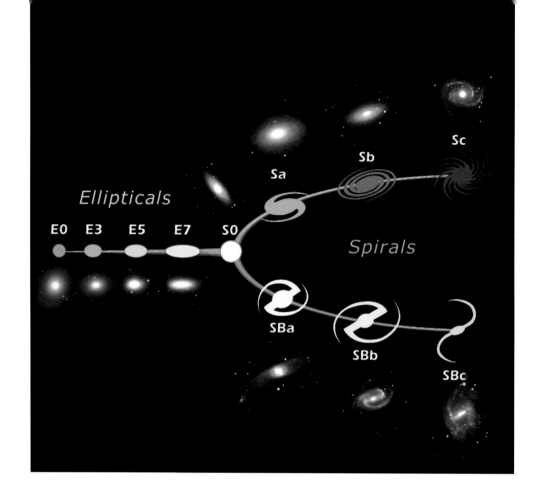

while spirals were called "late-type." This evolutionary picture appears to be supported by the fact that the disks of spiral galaxies are observed to be home to many young stars and regions of active star formation, while elliptical galaxies are composed of predominantly older stellar populations.

In fact, current evidence points to just the opposite: the early universe appears to have been dominated by spiral and irregular galaxies. In the currently favored picture of galaxy formation, called the *merger hypothesis*, present-day ellipticals formed as a result of mergers between these earlier building blocks.

Astronomers still use Hubble's galaxy classification today, although the initial concept is now seen as an oversimplification. Galaxy classification is a far more complex process than Hubble imagined, involving not only what our eyes can see in the visual part of the spectrum but also the conditions of a galaxy's initial collapse, its composition, and especially its evolution.

Spiral Galaxies

As Hubble's classification shows, spiral galaxies come in different shapes, but they do share a common structure. They all have a disk with stars, gas and dust, and a thick spherical central concentration of stars with little gas or dust, known as the bulge. The

> *"In the currently favored picture of galaxy formation, called the merger hypothesis, present-day ellipticals formed as a result of mergers between these earlier building blocks."*

whole structure is surrounded by a much fainter round halo of old stars, some of which reside in globular clusters. The range of masses for spiral galaxies is from about one to 1,000 billion solar masses, although the typical mass is about 100 billion solar masses. Typically, the size of the visible disk in a spiral galaxy goes from 15 to 300 thousand light-years in diameter — too large to really fathom — and a thickness of only one fiftieth of the diameter. Remarkably thin, rather similar to a plain dinner plate.

The whole system rotates around an axis joining the galactic poles. Although the inner part rotates like a solid body, the outer disk structure has a differential rotation pattern.

More than half of all observed galaxies are spirals. They are mostly found isolated in low density regions and are rare in the centers of galaxy clusters.

Two or more arms extend from the bulge, enclosing it. There are also several active star-forming regions, with hot glowing clouds of gas and dust forming the "stellar nurseries" that we see as nebulae outlining the spiral arms. The many hot young blue and white-blue stars make the structure of the arms very prominent. About half of all spirals show an additional component in the form of a central bar extending from the bulge, at the ends of which the spiral arms begin.

The nucleus of a spiral galaxy is typically much redder and smoother than its outer parts, often resembling elliptical galaxies; this indicates the presence of many old stars and not many star-forming regions, in contrast to the activity in the arms.

The space between the stars in these galaxies is filled with gas and dust, creating bright and dark patches and lanes along the spiral arms. In the Milky Way, seen from our viewpoint on Earth, the center itself is hidden from our view by thick layers of gas and dust.

It is thought that most spiral galaxies, if not all, harbor a supermassive black hole at the center, with a mass between 100,000 and 10 billion solar masses. Such black holes have

SUPERMASSIVE BLACK HOLE IN NEARBY GALAXY NGC 4438

Like the Milky Way, NGC 4438 also hosts a supermassive black hole. It is gorging itself on the banquet of material swirling around it in an accretion disk. Some of this material is spewed out from the disk in opposite directions. Acting like high-powered garden hoses, two jets of matter sweep out material in their paths. The jets eventually slam into a wall of dense, slow-moving gas, which produces the glowing material.

never been observed directly, but abundant indirect evidence for them exists. A 4-million solar mass supermassive black hole is located at the center of the Milky Way.

Most of the stars in a spiral galaxy are located either in the disk, following more or less circular orbits around the center, or in a spherical bulge around the galactic core. However, a number of stars inhabit a more diffuse spherical halo, either isolated or collected in globular clusters. The orbits of these stars are under close scrutiny by astronomers, but they may be unlike those of most stars in the Milky Way. Some orbit the center in the opposite direction or with a highly inclined or irregular orbit. It may very well be that these halo stars have been "acquired" over time from small galaxies that merged with the Milky Way.

The motion of halo stars does bring them through the disk on occasion. A number of small red dwarf stars close to the Sun are, for instance, thought to belong to the galactic halo, such as Kapteyn's Star and Groombridge 1830. Due to their irregular movement around the center of our galaxy these stars often move relatively fast when measured year by year relative to other stars.

> *"The origin of the spiral structure has been a matter for vigorous discussion among scientists for decades."*

The origin of the spiral structure has been a matter for vigorous discussion among scientists for decades. Soon after the discovery of the nature of spiral galaxies, scientists realized that stars could not permanently stay arranged in a spiral structure. Since the rotation speed of the galaxy varies with distance from the center of the galaxy, the arms would quickly become more and more curved, tightly embracing the core after only few rotations of the galaxy. Since we do not observe this effect, there has to be another explanation.

Today's most accepted explanation, the density wave theory, describes the arms as regions of enhanced density rotating in the disk of the galaxy. According to this theory, the number of stars in the spiral arms is not very different from the number of stars in the gaps between the arms. The often-used analogy for the density wave theory is a traffic jam on a highway, where cars enter and leave the jam, but the jam itself persists. In a spiral galaxy the arms have stars entering and leaving, but the arms themselves remain. The density wave rotates more slowly than the material in the galactic disk, so that stars and gas are able to "overtake" the wave. As gas in the interstellar medium passes into the density wave, it becomes denser, leading to the formation of new stars.

The hottest and brightest new stars burn their nuclear fuel at a high rate, so they have very short lifespans, living and dying very close to the density wave. This is why the spiral arms are traced out by the brightest stars, and in the gaps between the arms there are many faint stars but few very bright ones, since the brightest stars have already died. Gas and dust clouds indicating star-forming regions can also be seen on the inside of the spiral arms, lying slightly behind the density wave that has compressed them and triggered an explosion of star birth.

THE BARRED SPIRAL NGC 1672

NGC 1672, visible from the southern hemisphere, is seen almost face-on and shows regions of intense star formation. The greatest concentrations of star formation are found in the so-called starburst regions near the ends of the galaxy's galactic bar, seen running diagonally through the image. NGC 1672 is a prototypical barred spiral galaxy and differs from normal spiral galaxies in that the spiral arms do not twist all the way into the center. Instead, they are attached to the two ends of a straight bar of stars extending across the nucleus.

" *Despite the smooth, featureless appearance and the slower motion of their stars, elliptical galaxies are quite exciting dynamical systems.* "

Elliptical Galaxies

Despite the smooth, featureless appearance and the slower motion of their stars, elliptical galaxies are quite exciting dynamical systems. They appear as as fuzzy elliptical patches of light, ranging from nearly spherical to highly flattened ellipsoids, diminishing smoothly in brightness from the center outwards and rarely with any true structure visible. They have little visible gas and dust compared with spirals. Stars in elliptical galaxies are old and follow orbits that are randomly oriented within the galaxy. It also looks as if all elliptical galaxies have supermassive black holes in their center, and the mass of these black holes is connected with the mass of the galaxy. Elliptical galaxies do not have disks around them, although some bulges of disk galaxies look similar to elliptical galaxies.

Elliptical galaxies are believed to make up approximately 10 to 15 percent of galaxies in the local universe. They are most often found close to the center of rich, regular galaxy clusters and are less common in the early universe than they are now.

Ellipticals display a great variation in mass, ranging from dwarfs with a mass of just a few million solar masses to over one trillion solar masses for giant ellipticals. Their size varies from a few thousand light-years to 300,000 light-years. The smallest of the elliptical galaxies, which are called dwarf ellipticals, may be only a little larger than the largest globular clusters. At the other extreme, giant elliptical galaxies such as Messier 87 are among the largest galaxies in the universe. This is a much larger range of sizes than is seen for spiral galaxies. Giant ellipticals often dominate a galaxy cluster.

MESSIER 81

A multi-wavelength view of Messier 81's system of stars, dust, gas clouds, and spiral arms winding all the way down into the nucleus. Though the galaxy is located 11.6 million light-years away, the Hubble Space Telescope's view is so sharp that it can resolve individual stars, along with open star clusters, globular star clusters, and even glowing regions of fluorescent gas. This image combines data from the Hubble Space Telescope, the Spitzer Space Telescope and the Galaxy Evolution Explorer (GALEX) missions.

In most elliptical galaxies the lack of gas or dust means there is hardly any fuel available to feed star formation, so these galaxies have no newborn stars and no hot, bright, massive stars. Rather, elliptical galaxies are composed of old, low-mass stars, giving them a yellow-reddish color.

As we saw above, the Hubble classification of elliptical galaxies ranges from E0 for those that are most spherical, to E7, which have a long, thin shape. It is now recognized that the majority of ellipticals are of intermediate thinness, and that the Hubble classifications are often a result of the angle at which the galaxy is observed.

GIANT ELLIPTICAL GALAXY ESO 325-G004

This image shows the diverse collection of galaxies in the cluster Abell S0740, over 450 million light-years away in the direction of the constellation of Centaurus. The giant elliptical galaxy ESO 325-G004 looms large at the cluster's center. The galaxy is as massive as 100 billion solar masses. Hubble has resolved thousands of globular star clusters orbiting ESO 325-G004, which appear as pinpoints of light contained within the diffuse halo. Other fuzzy elliptical galaxies dot the image. Some have evidence of a disk structure that gives them a bow-tie shape. Several spiral galaxies are also seen.

NGC 7049

Tightly wound, almost concentric arms of dark dust encircle the bright nucleus of the galaxy NGC 7049. This galaxy is on the border between ellipticals and spirals (an S0 or Sa type).

MESSIER 87

The monstrous elliptical galaxy Messier 87 is the home of several trillion stars, a supermassive black hole, and a family of 13,000 globular star clusters. Messier 87 is the giant elliptical galaxy at the center of our neighboring Virgo Cluster of galaxies, which contains some 2,000 galaxies. The jet is a black-hole-powered stream of material that is being ejected from the core of the galaxy. The 120,000-light-year-diameter galaxy lies at a distance of 54 million light-years from the Sun in the spring constellation of Virgo.

> *"The Atlas of Peculiar Galaxies was a pioneering attempt to solve the mystery of the bizarre shapes of galaxies observed."*

IRREGULAR GALAXY NGC 4449

Hundreds of thousands of vibrant blue and red stars are visible in this image of galaxy NGC 4449. Hot bluish-white clusters of massive stars are scattered throughout the galaxy, interspersed with numerous dustier reddish regions of current star formation. Massive dark clouds of gas and dust are silhouetted against the flaming starlight. NGC 4449 has been forming stars for the past several billion years, but is currently forming stars at a much higher rate than in the past. At the current rate, the gas supply that feeds stellar production will only last for another billion years or so.

Irregular and Peculiar Galaxies

A number of observed galaxies do not find a place in the Hubble sequence and cannot really be classified as either elliptical or spiral. These are named irregular galaxies, and have little symmetry in their structure, or do not have a structure at all. They do not have a bulge, and although they sometimes appear to have the remains of a structure (irregular type I), it is not clear enough to really fit them into the Hubble sequence. Irregular type IIs have no trace of a structure.

Irregular galaxies have few systematic features. They vary in mass from 100,000 to 10 million solar masses and in size from a few thousand to several tens of thousands of light-years.

In the early days of Hubble's original classification, it was common to classify all galaxies that were neither spiral nor elliptical as irregular, but it is more common today to make a further distinction between "normal" irregular galaxies and peculiar galaxies, which have an unusual shape, size, structure, or composition. Most of these are objects that have been tidally distorted (see page 85) by interaction with another galaxy. They often appear in pairs of distorted members, with tails of stars and gas that have been flung out from the main body of the galaxy.

In the 1960s the American astronomer Halton Arp pointed the world famous 5-meter Palomar telescope toward 338 of these peculiar, interacting galaxies, capturing some of the most memorable astronomical images in history. His remarkable catalog, published as the *Atlas of Peculiar Galaxies*, was a pioneering attempt to solve the mystery of the bizarre shapes of galaxies observed by ground-based telescopes (read more about Arp in Chapter 3).

Today, the peculiarity of the galaxies seen by Arp and others is well understood, as the structures created by gravity's destructive dance when galaxies collide and merge.

> *"Astronomers are now convinced that the galaxy collisions play a fundamental role in establishing the many different galaxy shapes."*

DWARF IRREGULAR I ZWICKY 18

I Zwicky 18 is classified as a dwarf irregular galaxy and is much smaller than our Milky Way. The concentrated bluish-white knots embedded in the heart of the galaxy are two major starburst regions where stars are forming at a furious rate. The wispy blue filaments surrounding the central starburst regions are bubbles of gas that have been blown away by stellar winds and supernova explosions from a previous generation of hot, young stars. This gas is now heated by intense ultraviolet radiation unleashed by hot, young stars. A companion galaxy lies just to the left of I Zwicky 18 and is likely interacting with I Zwicky 18 by gravitationally tugging on the galaxy. The interaction may have triggered the galaxy's recent star formation that is responsible for the youthful appearance. The reddish extended objects surrounding I Zwicky 18 and its companion are ancient, fully formed galaxies of different shapes that are much farther away.

Irregular galaxies are not an exception in the universe. It is thought that about one fifth of all galaxies are either irregular or peculiar in some way.

The study of these types of galaxies has gradually put more emphasis on the important evolutionary influence of interactions between galaxies. Astronomers are now convinced that this process plays a fundamental role in establishing the many different galaxy shapes. These interactions are also important in stimulating the formation of stars and star clusters.

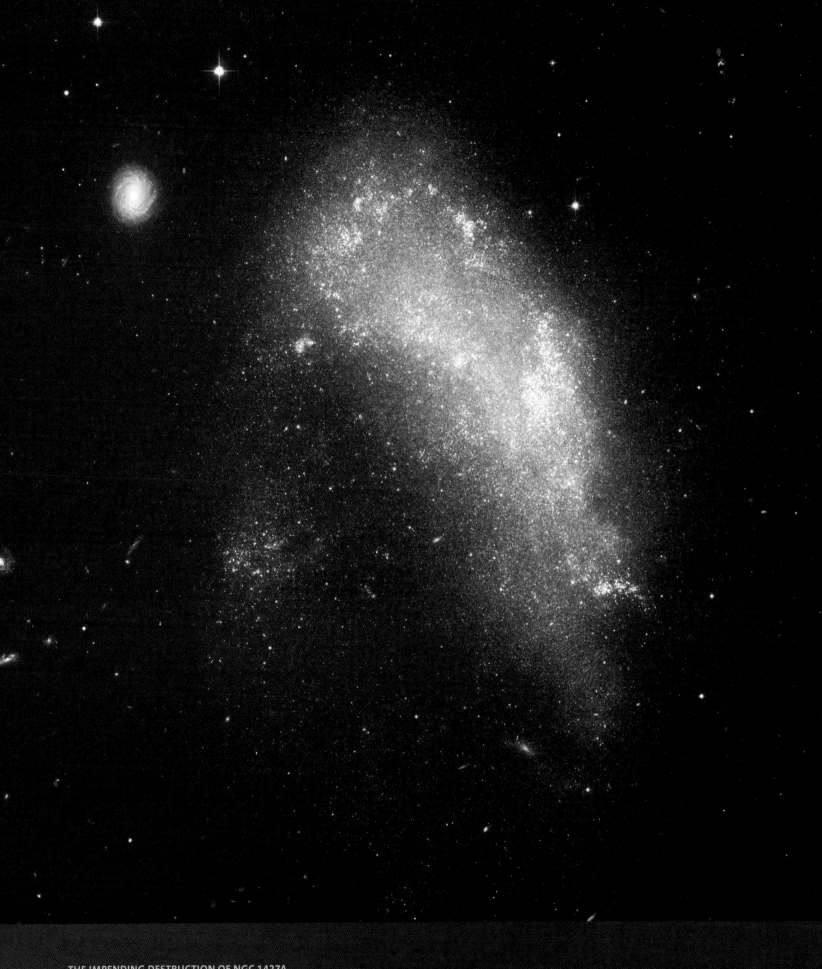

THE IMPENDING DESTRUCTION OF NGC 1427A

The irregular galaxy NGC 1427A is a spectacular example of an imminent stellar battle between galaxies. Under the gravitational grasp of a large cluster of galaxies, called the Fornax cluster, the small bluish galaxy is plunging headlong into the group at more than two million kilometers per hour.

2 HOW DO GALAXIES FORM AND EVOLVE?

GALAXIES, GALAXIES EVERYWHERE

The Hubble Ultra Deep Field is one of the most important deep images of recent years. Hubble has imaged nearly 1,000 galaxies in this so far deepest visible-light image of the cosmos. This galaxy-studded view represents a "deep" core sample of the universe, cutting across billions of light-years, showing galaxies of various ages, sizes, shapes, and colors. The smallest, reddest galaxies may be among the most distant known, existing when the universe was just 800 million years old. The nearest galaxies — the larger, brighter, well-defined spirals and ellipticals — are seen as they were about one billion years ago.

The evolution of galaxies is the ultimate chicken-and-egg situation. Did the shapeless matter first gravitate together in larger structures, giving birth to stars and then to supermassive black holes? Or did the black holes form first and then trigger the creation of the first generation of stars in the centers of galaxies?

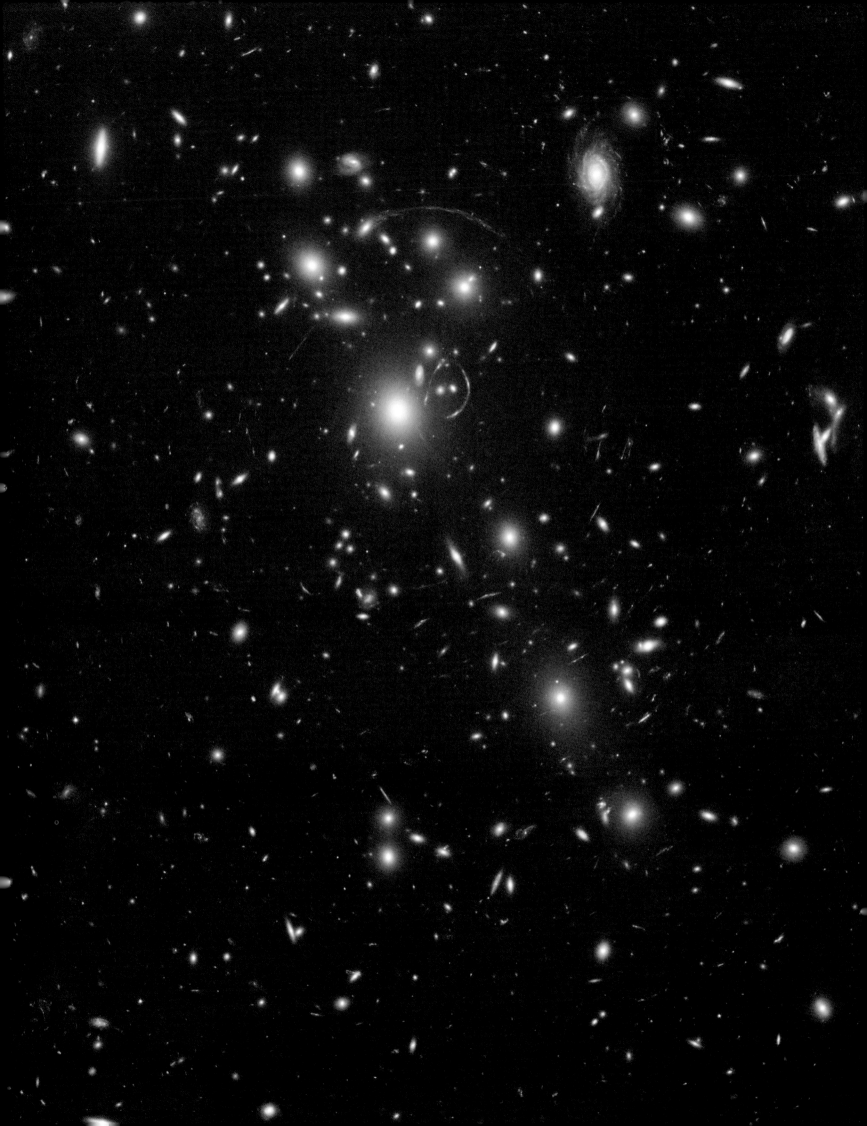

" Galaxy clusters are the largest gravitationally bound objects we know. "

The formation and evolution of galaxies is one of the most hotly debated topics in modern astronomy. Why do we see different types of galaxies, and how do the supermassive black holes that are found in all, or nearly all, galaxies fit into the equation?

Important clues to the origin of galaxies are found by looking at surveys of large areas of the sky. Some regions are more densely populated than average, with galaxies found in small groups, or in large crowds of thousands of galaxies, called galaxy clusters. Clusters are often grouped in superclusters and even larger structures that extend across large swathes of the mappable universe.

Galaxy clusters are the largest gravitationally bound objects we know and have a well-populated central core and a spherical shape. Typically, their sizes range between 5 and 30 million light-years, while their mass is of the order of one million billion solar masses. Compared to the fields outside the clusters, the cluster centers are populated almost solely by elliptical and lenticular (or S0; on the border between ellipticals and spirals) galaxies, with hardly any ongoing star formation. So, there is a clear connection between the environment and the galaxy types found. This relationship makes many scientists believe that spirals were once numerous in clusters but have been transformed into elliptical or lenticular systems via galaxy mergers.

ABELL 1703, A MASSIVE GALAXY CLUSTER

Located in the northern celestial hemisphere, Abell 1703 is composed of hundreds of galaxies, here seen in yellow. Most of these are elliptical galaxies. The cluster galaxies act as a powerful cosmic telescope, or gravitational lens, that bends and stretches the light from more distant galaxies (many of which are spirals). In the process it distorts their shapes and produces multiple banana-shaped images of the original galaxies. Abell 1703 is located 3 billion light-years from Earth.

Looking Back in Time

A single human lifetime, or even the lifetime of the entire species, is far too short to observe the evolution of a galaxy. But the speed of light comes to our rescue in a curious way. It is a very high speed indeed — roughly 300,000 kilometers per second — but it is still finite. Galaxies are millions or even billions of light-years (the distance traveled by light in a year) away from Earth. Due to the finite speed of light, the more distant an object is, the longer the travel time of the light to Earth, and so the further into the past we observe it once the light arrives. Observing a distant galaxy is like traveling back in time. This gives us the ability to study the changes in galaxies over time by observing them at different distances, and thus at different epochs.

A SPIRAL GALAXY BEING TRANSFORMED INTO AN ELLIPTICAL?

NGC 4522 is a spectacular example of a spiral galaxy currently being stripped of its gas content (upwards in the picture). The galaxy is part of the Virgo galaxy cluster, and its rapid motion within the cluster results in strong "winds" across the galaxy ripping out parts of its gas. Scientists estimate that the galaxy is moving at more than 10 million kilometers per hour. A number of newly formed star clusters are being ripped out from the disk of the galaxy and can be seen in the Hubble image. Gas stripping is believed to significantly influence the shape and evolution of spiral galaxies in galaxy clusters. This stripped spiral galaxy is located some 100 million light-years away from Earth and may one day end up as an elliptical galaxy.

" The first galaxies formed within the first five percent of the universe's lifetime. Compared to an average human lifespan the universe had not yet reached school age."

In the Beginning...

The Big Bang was the event that brought about the origin of space and time and the laws of nature and of all matter and energy. During the first few hundred thousand years after the Big Bang the universe was remarkably homogeneous and far too hot for chemical elements to form. The universe consisted of a soup of subatomic particles and radiation.

Around 380,000 years after the Big Bang, as the universe cooled to 3,000°C, the first complete hydrogen and helium atoms began to form. Images of the microwave background radiation, emitted around this time, indicate that the seemingly featureless sea of cosmic particles and radiation already then showed signs of structure. In the most accepted picture of galaxy formation, these subtle variations, likely dominated by dark matter (see box on page 47), in an otherwise smooth universe were the seeds that grew to form the first galaxies. These first primordial structures gravitationally attracted gas and dark matter to the densest areas, where the first proto-galaxies formed and later grew into galaxies.

At this point in time the "normal matter" in the universe was composed almost exclusively of hydrogen and helium (although this is only four percent of the full story; see the box: Dark matter.) From these gas clouds the first stars emerged and the first visible proto-galaxies were formed, ending what is known as the "Dark Ages."

The earliest galaxies observed so far date back to roughly 750 million years after the Big Bang, but earlier galaxies may well exist. Although 750 million years may seem like a long time, when compared with the universe's age of 13.7 billion years, the first galaxies formed within the first five percent of the universe's lifetime. Compared to an average human lifespan the universe had not yet reached school age. A lot happened in the universe's infancy!

> *"The creation of the supermassive black hole appears to play a key role in actively regulating the growth of a galaxy."*

Evolution of Galaxies

The conventional picture of galaxy formation, the merger hypothesis, explains that protogalaxies grew and evolved in the early universe by the "fusion" of smaller building block galaxies. Large sky surveys have shown us that galaxies in the nearby and "half-distant" universe seem to gather in large filaments, creating a great cosmic web. The intersections where the filaments meet are home to the densest clusters of galaxies.

We also know that within a billion years of a galaxy's formation, key structures begin be visible. Globular clusters, the central supermassive black hole, and a galactic bulge of stars form. The creation of the supermassive black hole appears to play a key role in actively regulating the growth of a galaxy. It grows by eating innocent stars that happen to come too close, but also influences the birth of new stars. Observations of distant galaxies also show that during this early epoch in the universe's life, galaxies undergo a major burst of star formation.

Despite its many successes, this conventional picture has its limitations. It cannot completely explain the variety of shapes we observe in galaxies, from the featureless round elliptical galaxies to the pancake-flat disk galaxies.

Evolution of Spiral Galaxies

One of the main challenges for the merger hypothesis is to explain the great number of spiral galaxies in the present universe — more than half of the observed galaxies. Spirals are fragile, and interactions with nearby galaxies quickly destroy their shape. It may be that the spirals we see have not yet interacted with other large galaxies, and that this is the reason they still have their beautiful structure and ongoing star formation.

During the first few billion years of a galaxy's life, the accumulated matter settles into a galactic disk. The galaxy continues to absorb infalling material from high velocity clouds

Dark matter — Now you see it, now you don't.

Dark energy — Now you understand it, now you don't.

Most of the light that comes to us from the universe is either the glow of stars, or came originally from stars and was altered on its way to us. But there is clear evidence that there is "stuff" that we can't see directly, and that this "dark matter" is the main constituent of galaxies and the universe as a whole.

The astronomer Fritz Zwicky was the first to notice the "missing mass" in 1933, observing the motions of the galaxies near the edges of the Coma Cluster. The measured motion of the galaxies in the cluster is too fast for them to be held together solely by the gravity of the visible stars comprising the galaxies.

Another "dark matter" milestone was the observation of rotation curves in spiral galaxies. In 1975, Vera Rubin discovered that the stars in spiral galaxies orbit with roughly the same speed, without the expected slower motion near the edge of the visible parts of the galaxy. This result showed that upwards of 50 percent of the mass of galaxies is contained in the relatively dark galactic halo.

The nature and composition of this unseen matter remain unknown. It constitutes a happy hunting ground for observers and theoreticians. Proposals have included brown dwarf stars and planets (collectively called MACHOs), quantum black holes produced in the early universe, clouds of dark gas, ordinary and heavy neutrinos, and a whole zoo of exotic particles, such as Weakly Interacting Massive Particles.

Results from the Wilkinson Microwave Anisotropy Probe and other telescopes lead to the somewhat discouraging conclusion that all of our technology and brainpower so far has only managed to tell us about the constituents of 4 percent of the cosmos. Almost a quarter of the rest of the universe is composed of dark matter (22 percent), while the remaining 74 percent is in the form of the even more elusive and mysterious dark energy, about which we know hardly anything. What we do know is that dark energy exerts a repulsive force on the universe that opposes the gravitational attraction between galaxies and therefore fits well with recent observations that the expansion of the universe is accelerating.

THE BULLET CLUSTER

Perhaps the most convincing evidence for dark matter is the Bullet Cluster. This is actually composed of two colliding clusters, many members of which can be seen in the background visible-light image. Overlaid on this is the distribution of the majority of the normal matter in red (a very hot X-ray emitting gas) and a map of the majority of the mass of the cluster in blue (measured from the effect of gravitational lensing). The offset between the red X-ray gas and the blue mass measurement shows an amazing difference between the normal and the dark matter in the two clusters. The red bullet-shaped clump on the right is the hot gas from one cluster, which passed through the hot gas from the other larger cluster during the collision. Both gas clouds were slowed by a drag force, similar to air resistance, during the collision. In contrast, the dark matter (in blue) was not slowed by the impact because it — apparently — does not interact directly with itself or the gas except through gravity. This result is dramatic direct evidence that most of the matter in the clusters is dark, and very different from normal matter!

> *"After the collision, the captured stars rock back and forth and form the faint shells just as water forms ripples when we toss a rock into a pond."*

and dwarf galaxies throughout its life. This matter is mostly hydrogen and helium. The cycle of stellar birth and death slowly increases the abundance of elements heavier than hydrogen and helium.

The theories for the formation of spiral galaxies include the clustering of dark matter halos. In the early universe, galaxies were composed mostly of gas and dark matter and not too many stars. By accreting smaller galaxies a spiral galaxy "wannabe" becomes more massive, while the dark matter is mostly concentrated in a halo in its outer parts. The gas, however, contracts relatively quickly and as a consequence begins to rotate faster, just as a spinning skater speeds up if she pulls in her arms. The final result is a thin and rapidly rotating disk galaxy.

Why the contraction stops is still a mystery, and computer simulations of disk galaxy formation have not yet been able to fully reproduce the rotation speed and size of these disk galaxies. It has been suggested that radiation from bright, newly formed stars, or the supermassive black hole in most, or all, galaxy centers, can regulate the process by slowing the contraction of a disk as it forms. It has also been suggested that the halo of the dark matter can exert a "drag" on the galaxy, thus stopping disk contraction. Maybe already at this stage galaxy collisions and mergers play an important role by feeding "new" gas stars and dark matter to galaxies.

Evolution of Elliptical Galaxies

The traditional portrait of elliptical galaxies paints them as galaxies where star formation has finished after the initial burst, leaving them to shine with only the light from their aging stars. In the outer regions, large elliptical galaxies typically have an extensive system of globular clusters.

Lack of form and an old stellar population initially led astronomers to believe that elliptical galaxies formed earlier than spirals. However, recent observations have found young,

A LARGE GALAXY IN THE MAKING

A large galaxy 10.6 billion light-years away from Earth is seen "stuffing" itself with smaller galaxies, caught like flies in a web of gravity. The galaxy is so far away that astronomers are seeing it as it looked in the early formative years of the universe, only two billion years after the Big Bang. This image is a strong confirmation of the most common theories for how galaxies evolve. MRC 1138-262, nicknamed the "Spiderweb Galaxy," has dozens of star-forming satellite galaxies that are in the process of merging.

blue star clusters inside a few elliptical galaxies along with other structures that can be explained by galactic collisions. The merger hypothesis predicts that an elliptical galaxy is the result of a long process where two galaxies of comparable mass, of any type collide and merge. Such major galactic mergers are thought to have been common in early times but may still occur, though less frequently today.

Observations reveal that roughly half of all ellipticals have faint shells that may be the result of head-on collisions with smaller galaxies. After the collision, the captured stars rock back and forth and form the faint shells just as water forms ripples when we toss a rock into a pond.

With the merger hypothesis, it is today widely accepted that the main driving force for the evolution of elliptical galaxies is the mergers of smaller galaxies. If two approaching galaxies are of similar size, the resultant merged galaxy will not resemble either of the two pre-merger galaxies. During the merger, stars and dark matter in each galaxy are

> *" The resulting galaxy is dominated by stars that orbit the center in a complex, and random, web of orbits, just the behavior we see in elliptical galaxies. "*

NGC 1132, A GIANT ELLIPTICAL GALAXY

NGC 1132 is a cosmic fossil — the aftermath of an enormous multi-galactic pile-up, where the carnage of collision after collision has built up a brilliant but fuzzy giant elliptical galaxy far outshining typical galaxies. NGC 1132, together with the small dwarf galaxies surrounding it, are dubbed a "fossil group," as they are most likely the remains of a group of galaxies that merged together in the recent past. NGC 1132 is seen surrounded by thousands of ancient globular clusters, swarming around the galaxy like bees around a hive. These globular clusters are likely to be the survivors of the disruption of their cannibalized parent galaxies that have been eaten by NGC 1132 and may reveal its merger history. In the background, there is a stunning tapestry of numerous galaxies that are much farther away.

affected by the approaching galaxy. Toward the late stages of the merger, the shape of the galaxy changes so quickly that stellar orbits are affected dramatically and no longer bear any relation to their original courses. If two spiral galaxies collide, their stars may initially be rotating in an orderly manner in the plane of the disk, but during the merger, this ordered motion is transformed into random motions. The resulting galaxy is dominated by stars that orbit the center in a complex, and random, web of orbits, just the behavior we see in elliptical galaxies.

Star formation was more pronounced during the mergers that formed most of the elliptical galaxies we see today, which likely occurred between one and ten billion years ago, when there was much more gas (and thus more molecular clouds) in galaxies. Away from the center of the galaxy, gas clouds ran into each other and produced shocks that stimulated the formation of new stars. The result of all this violence is that ellipticals tend to have very little gas available to form new stars after they merge.

GALAXY COLLISIONS

NGC 3256

NGC 3256 is an impressive example of a peculiar galaxy. The telltale signs of the collision are two extended luminous tails swirling out from the galaxy. NGC 3256 belongs to the Hydra-Centaurus supercluster complex and provides a nearby template for studying the properties of young star clusters in tidal tails. The system hides a double nucleus and a tangle of dust lanes in the central region. The tails are studded with a particularly high density of star clusters.

Like majestic ships in the grandest night, galaxies can slip ever closer until their mutual gravitational interaction begins to mold them into intricate figures that are finally, and irreversibly, woven together. It is an immense cosmic dance, choreographed by gravity.

> *"Our usual impression of the starry night sky is that of a nearly motionless dome. A single human life span is just the blink of an eye on a cosmological time scale."*

When two or more galaxies collide, it's not like marbles hitting each other. The individual components of galaxies — stars, dust, gas, dark matter — are spread widely and may pass unharmed through the collision. At worst, gravity will fling them outwards, creating long streamers that stretch for a hundred thousand light-years or even more.

The colliding galaxies themselves, trapped in their deadly gravitational embrace, will continue to orbit each other, ripping out gas, stars, and creating trails of matter in the cosmos. Eventually, hundreds of millions of years later, the galaxies involved in the collision will settle into one single, combined galaxy. It is believed that most if not all present-day galaxies, including the Milky Way, were formed from such a coalescence of smaller galaxies, occurring over billions of years.

Cosmologists think that this is how galaxies grow, through an intricate process of continuous mergers. Galaxies grow bigger by devouring smaller ones, dwarf galaxies being the favorite meals of larger galaxies.

More common in the early universe than they are today, galaxy mergers are thought to be a primary driving force of cosmic evolution, sparking frenetic births and explosive stellar deaths. Our own Milky Way contains the debris of the many smaller galaxies it has encountered and devoured over billions of years. But even our galaxy is not at the top of the food chain (see Chapter 5).

The First Observations

Our usual impression of the starry night sky is that of a nearly motionless dome. A single human life span is just the blink of an eye on a cosmological time scale. In reality, the universe is like a churning cauldron, in constant movement, but we would need to watch it for vastly longer than a lifetime to perceive the motions of most stars and galaxies.

ARP 81

Arp 81 is a strongly interacting pair of galaxies, seen about 100 million years after their closest approach. It consists of NGC 6621 (to the right) and NGC 6622 (to the left). NGC 6621 is the larger of the two and is a very disturbed spiral galaxy. The encounter has pulled a long tail out of NGC 6621 that has now wrapped behind its body. The collision has also triggered extensive star formation between the two galaxies. Scientists believe that Arp 81 has a richer collection of young massive star clusters than the notable Antennae galaxies (which are much closer than Arp 81). The pair is located in the constellation of Draco, approximately 300 million light-years away from Earth.

Given enough time, we would see stars and galaxies move. Stars orbit the center of the Milky Way, and galaxies are pulled together by each other's gravity, colliding and eventually generating "new" galaxies.

The first recorded galaxy interactions go back before we knew that galaxies were made of stars, dust, and dark matter. William Parsons, the third Earl of Rosse, observed the heavens in Ireland with his famous 1.8-meter (72-inch) telescope, called the "Leviathan of Parsonstown." A drawing he made in 1845 shows not only the clear spiral structure of the Whirlpool Galaxy (Messier 51) but one of the spiral arms connecting to its companion galaxy. Comparing this drawing with a recent Hubble image shows a remarkable resemblance. Not only the placement of the connection arm between the main galaxy and the companion are in beautiful agreement with reality, but the dark dust patches stand out where the main galaxy crosses in front of the companion.

> *" During the early part of the Second World War, Holmberg devised an extraordinarily clever experiment in a darkened barn in Sweden. "*

The First Simulations

With the advent of photography and the increasing size of telescopes, more and more interacting systems could be recognized, and this naturally attracted the interest of theoreticians. Around 1940 the Swedish astronomer Erik Holmberg was cataloging close pairs of galaxies and began to wonder what would happen to galaxies as they approached each other. He showed that galaxies often appear in groups and pairs, and he also realized that it would be possible, using statistics, to determine the masses of pairs of galaxies knowing the motions of the components. The observations also resulted in the discovery that satellite galaxies often move in certain orbits, which is known as the "Holmberg effect."

During the early part of the Second World War, Holmberg devised an extraordinarily clever experiment in a darkened barn in Sweden. He used light bulbs to represent the stellar distribution in a galaxy, and the intensity of the light simulated gravitational strength. The gravitational forces were then measured by photocells, and so Holmberg could, step by step, and with care and patience, follow events as two galaxies passed close to each other. He was able to show that such collisions can make galaxies spiral closer together and merge, and even found the rudiments of the tidal spiral structures, bridges, and tails that are so familiar today. In 1941 Holmberg published the results of this ingenious analog simulation — probably his most significant work on the effects of interacting galaxies.

Holmberg's experiment was so ahead of its time that his results were not confirmed until about 30 years later. In the meantime, observations of colliding galaxies had begun to show an amazing variety of forms. Starting from the Palomar Sky Survey, Boris A. Vorontsov-Velyaminov in Moscow cataloged interacting galaxies and other oddities and published his first catalog, *Atlas and Catalogue of Interacting Galaxies*, Part 1, in 1959. This was a limited photographic printing in Russian and was not distributed as widely as it deserved, especially in the West.

THE FIRST RECORDED GALAXY COLLISION

Messier 51 as seen by William Parson in 1845 (top) and the Hubble Space Telescope (bottom). William Parson's drawing is the first known depiction of interacting galaxies — long before people even knew what galaxies really were.

THE FIRST SIMULATION OF MERGING GALAXIES

One of Holmberg's pioneering figures from 1941 is shown here. He devised a clever experiment, carried out in a darkened barn in wartime Sweden, using light bulbs to represent stars.

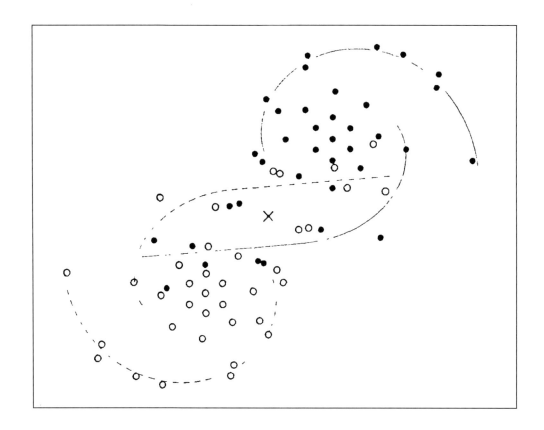

Following this work and adding other galaxies as he became aware of them, the astronomer Halton Arp in 1962 turned the great eye of Palomar on the 338 peculiar and interacting galaxies illustrated in his famous 1966 atlas, *The Atlas of Peculiar Galaxies*, which still represents some of the finest galaxy images ever obtained photographically. Some of his galaxy pairs were not only distorted but showed connecting filaments of stars that challenged the contemporary intuitive understanding of the workings of gravity.

Arp compiled the catalog in a pioneering attempt to solve the mystery of the bizarre shapes of galaxies observed. He also worked with Barry Madore, compiling a more comprehensive survey of the southern sky, published in 1987. Today, the peculiarity of the galaxy structures seen by Arp and others is now well understood, created by gravity's partly destructive, partly creative, dance as galaxies collide.

Beautiful digital scans of the old photographic plates can be examined online at NASA's Extragalactic Database. (See the Resources section in the back of this book.)

Real computer simulations of merging galaxies were pioneered by two Estonian brothers, Alar and Juri Toomre, in 1972. They demonstrated that collisions are far more common than previously thought and that pairs of galaxies approach each other at very high

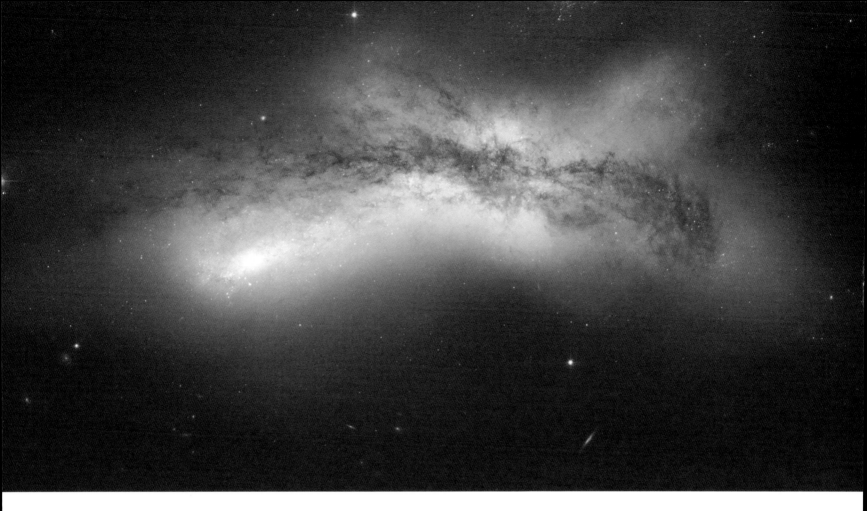

DUST LANES IN NGC 520

The contorted dust lanes in NGC 520, or Arp 157, are the product of a collision between two disk galaxies that started 300 million years ago. It exemplifies the middle stages of the merging process: the disks of the parent galaxies have merged together, but the nuclei have not yet coalesced. It features an odd-looking tail of stars and a prominent dust lane that runs across the center of the image and obscures the galaxy. NGC 520 is one of the brightest galaxy pairs in the sky and can be observed with a small telescope toward the constellation of Pisces, the Fish, having the appearance of a comet. It is about 100 million light-years away and about 100,000 light-years across.

velocities — hundreds of thousands of kilometers per hour. But because the distances between the stars are so large, very few actual collisions between stars occur. Most of the interaction as galaxies merge is gravitational, the pull of the giant masses generating immense tidal forces across the intermingling stars.

The Toomre brothers used computer techniques — quaintly primitive compared to modern ones — to model the Whirlpool Galaxy and some of the most puzzling of Arp's pairs — the Mice (page 87) and the Antennae galaxies (cover image). Most astrophysicists found that their intuition about how gravity could influence galaxies had led them astray, because galaxies contain matter spread over a wide area and thus respond very differently to a disturbance. The amount of "damage" is highly dependent on the relative

EARLY COMPUTER SIMULATION

An early simulation from the Estonian Toomre brothers showing how they modeled the Whirlpool Galaxy.

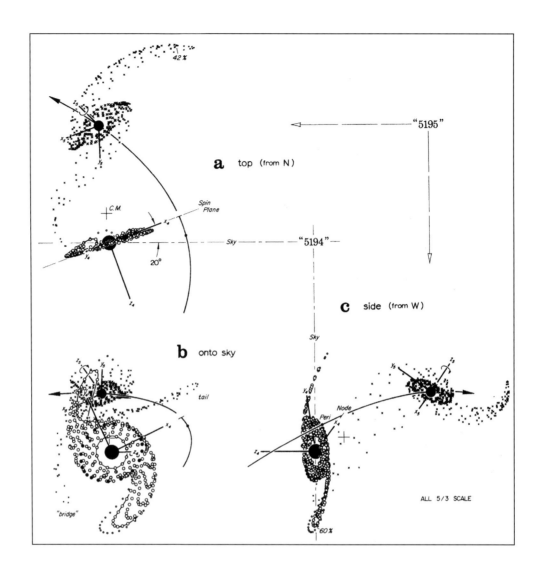

velocities of the two galaxies, the angle of incidence, their spin velocity, spin direction and more.

Because of the limitations of computers at that time, the Toomres had to make a number of simplifications. They concentrated the mass in the center of each galaxy, made each computed point stand in for a billion stars, etc. They still managed to build an impressive case explaining the counterintuitive tidal features. As computer simulations have evolved along with the technology, they have become more and more realistic, matching not only the shapes of galaxy pairs but their observed internal motions and tracing stars, cold gas, and unseen dark matter independently. The number of particles and the level of sophistication reach new heights every year.

> *"A major aspect of this extra star formation in colliding galaxies was only properly revealed when Hubble's sharp eyes were turned towards selected systems."*

Star Birth

From observations of galactic collisions it is clear that not only the shapes of the characters involved change; their colors, brightnesses, and temperatures are also modified during the interaction. Observations in the infrared band show that some interacting galaxies host the most extreme levels of star formation we can find anywhere in today's universe. The violent forces during a merger can trigger some galaxies to make thousands of solar masses of new stars each year. In the Milky Way only about one new star forms per year on average. Although stars almost never get close enough to actually collide in galaxy mergers, giant molecular clouds do collide with other molecular clouds. These collisions then induce the clouds to condense into new stars.

Pairs or groups of interacting galaxies are in general bluer than isolated galaxies. Furthermore they are bluer in a way that can be accounted for only if they have increased star formation. The U.S.–U.K.–Dutch Infrared Astronomical Satellite (IRAS) detected a handful of galaxies that were extraordinarily powerful emitters of infrared radiation. Some of these — named ultraluminous infrared galaxies (ULIRGs) — emit a trillion times as much energy as the Sun, almost all of it in the deep infrared. Among the first such "IRAS galaxies" identified were Arp 220 (page 62) and NGC 6240 (page 63), two systems with all the structural signs of being galaxy mergers viewed close to the final act as the nuclei of the two progenitors spiral together. Wider-ranging surveys revealed that almost all ULIRGs are in fact strongly interacting or merging galaxies. The enormous burst of star formation that generates the energy to heat it must somehow be triggered by the galaxy collision.

A major aspect of this extra star formation in colliding galaxies was only properly revealed when Hubble's sharp eyes were turned towards selected systems. Among the observatory's first scientific discoveries was that galaxies with very active star formation, particularly those in colliding and merging systems, contain large numbers of super star

ARP 220, ONE OF THE FIRST ULTRALUMINOUS INFRARED GALAXIES FOUND

Arp 220 appears to be an odd-looking single galaxy but is in fact a nearby example of the aftermath of a collision between two spiral galaxies. It is the brightest of the three galactic mergers closest to Earth, about 250 million light-years away in the constellation of Serpens, the Serpent. The collision, which began about 700 million years ago, has sparked a giant burst of star formation, resulting in about 200 huge star clusters in a packed, dusty region of about 5,000 light-years across (about 5 percent of the Milky Way's diameter). The amount of gas in this tiny region equals the amount of gas in the entire Milky Way Galaxy. The star clusters are the bluish-white bright knots visible in the Hubble image. Arp 220 glows brightest in infrared light and was one of the first so-called ultraluminous infrared galaxies found. Previous Hubble observations, taken in the infrared at a wavelength that looks through the dust, have uncovered the cores of the parent galaxies almost melted together, only 1,200 light-years apart.

NGC 6240, LAST STAGES OF A COLLISION

NGC 6240 is a peculiar, butterfly, or perhaps lobster-shaped, galaxy consisting of two smaller merging galaxies. It lies in the constellation of Ophiuchus, the Serpent Holder, some 400 million light-years away. Observations with NASA's Chandra X-ray Observatory have disclosed two giant black holes, only about 3,000 light-years apart, which will drift toward one another and eventually merge together into a larger black hole. The merging process, which began about 30 million years ago, triggered dramatic star formation and sparked numerous supernova explosions. The merger will be complete in some tens to hundreds of millions of years. The Hubble data were supplemented with infrared data from the Spitzer Space Telescope.

> *"A supermassive black hole located at a galaxy's core feeds on the gas and disrupted stars coming from the surrounding galaxy."*

clusters — clusters more compact and richer in young stars than we are used to seeing in our neighborhood. Today there is a particular focus on finding out what these massive star clusters will someday turn into. They look much as we would have expected the old globular clusters in the Milky Way and other galaxies to have appeared when they were very young.

If elliptical galaxies indeed turn into globular clusters, it would eliminate an important problem with the merger hypothesis. Elliptical galaxies often have a larger number of globular clusters containing old stars that would not easily survive a merging event. However, if the mergers preferentially create stars in super star clusters, and if these eventually evolve to look like the globulars, the merger hypothesis seems to explain things well.

Feeding Black Holes

A supermassive black hole located at a galaxy's core feeds on the gas and disrupted stars coming from the surrounding galaxy. As the material falls inwards it will usually form into a spinning disk orbiting the black hole. A small portion of the infalling material never reaches the black hole but is spun up by the surrounding magnetic fields and ejected at nearly the speed of light as oppositely directed jets (see NGC 4438 on page 26) — often radiating copious amounts of radio radiation. If one of the jets points roughly towards us, we will have a relatively clear view of the regions very close to the black hole, where much of the light is emitted, and we see a quasar. If we see the disk edge-on, our view of the black hole action may be obscured by the material in the disk. It is only in the last couple of decades that astronomers have begun to appreciate how the many differing types of so-called active galaxies can be understood simply in terms of how their appearance varies depending on which direction we view them from.

There have long been hints that galaxy collisions make conditions more favorable for the occurrence of active galaxies. Possibly the most compelling connection appeared in

MAGNIFICENT MESSIER 82

The ongoing violent star formation due to an ancient encounter with its large galactic neighbor, Messier 81, gives Messier 82 its disturbed appearance. The huge lanes of dust that crisscross Messier 82's disk are another telltale sign of the flurry of star formation. The numerous hot new stars emit not only radiation but also particles called a stellar wind. Stellar winds streaming from these hot new stars have also combined to form a fierce galactic superwind. This superwind compresses enough gas to make millions of more stars and blasts out of towering plumes of hot ionized hydrogen gas, above and below the disk of the galaxy (seen the red in the image). Most of the pale objects sprinkled around the main body of M82 that look like fuzzy stars are actually massive super-star clusters about 20 light-years across and containing up to a million stars.

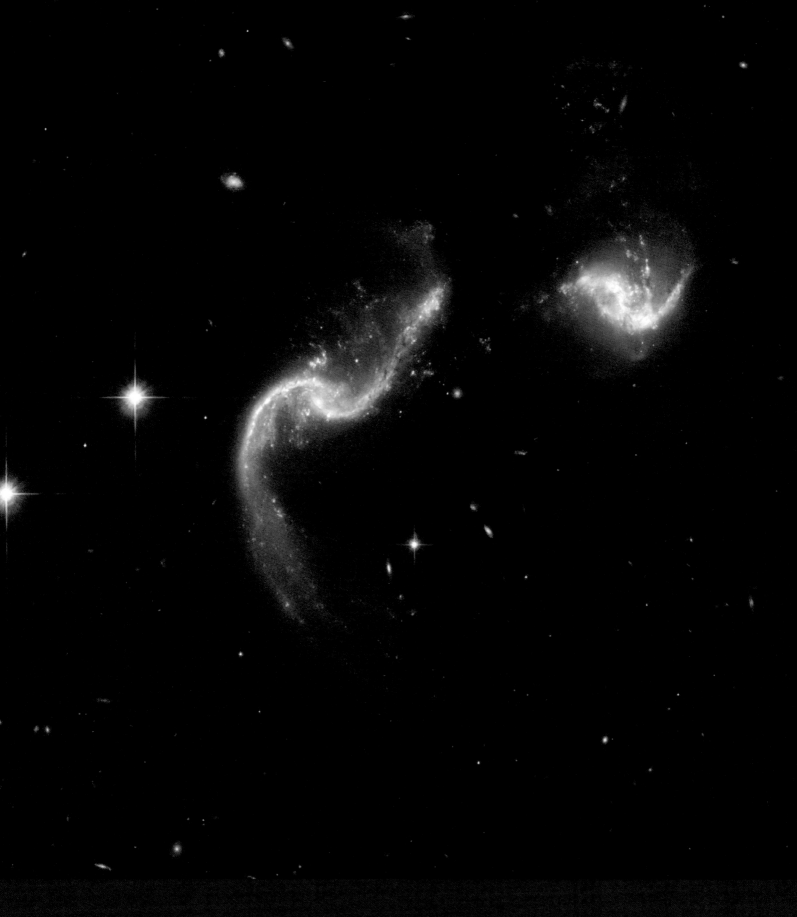

FIREWORKS OF NEWBORN STARS IN ARP 256

Arp 256 is a stunning system of two spiral galaxies in an early stage of merging. The Hubble image displays two galaxies with strongly disrupted shapes and an astonishing number of blue knots of star formation that look like exploding fireworks. The galaxy to the left has two extended ribbon-like tails of gas, dust, and stars. The system is a luminous infrared system radiating more than a hundred billion times the luminosity of our Sun. Arp 256 is located in the constellation of Cetus, the Whale, about 350 million light-years away.

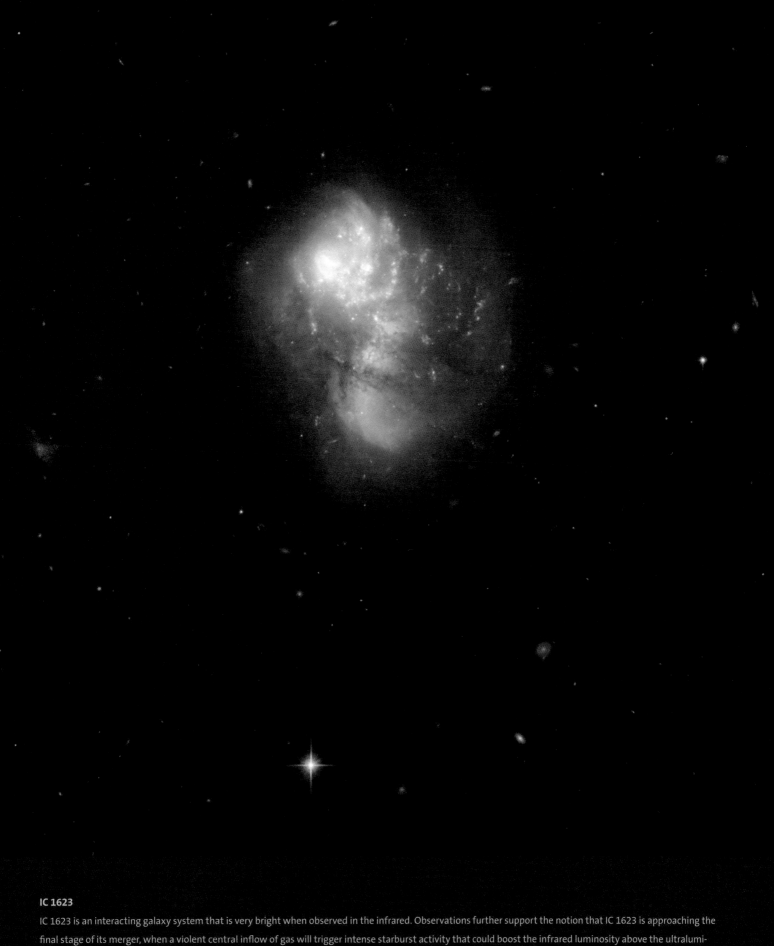

IC 1623

IC 1623 is an interacting galaxy system that is very bright when observed in the infrared. Observations further support the notion that IC 1623 is approaching the final stage of its merger, when a violent central inflow of gas will trigger intense starburst activity that could boost the infrared luminosity above the ultraluminous threshold. The system will likely evolve into a compact starburst system similar to Arp 220. IC 1623 is located about 300 million light-years away from Earth.

QUASAR HOST GALAXY

Hubble has captured a quasar together with its host galaxy that is usually drowned out in the glare from the quasar. Most of the host galaxies observed are distorted, a clear sign that the systems have been caught in the act of merging. Here, at least five shells of stars surrounding the quasar (center) can be seen. The inescapable conclusion is that galaxy collisions can trigger activity in the black holes in the centers of active galaxies.

Hubble images of the galaxies around quasars (page 68). These galaxies are usually lost in the brilliant and blurred light of the cores as seen from the ground. Seen with Hubble many of these galaxies showed not just interacting companions but a particular type — very small companion galaxies lying close to or within the quasar galaxy itself. Such a configuration is theoretically very effective at channeling stars and gas inward and outward in the victim galaxy and is one way that a formerly dormant black hole could be roused to an episode of quasar emission. Gas is fed to the huge black hole in the center of a massive galaxy, activating the nucleus into releasing unimaginable amounts of energy.

There is further evidence that a supermassive black hole appears to be able to regulate the growth of galaxies through some kind of — so far — uncertain feedback mechanism. It may be that the black hole feeds on the surrounding galactic material, producing enormous amounts of energy (expelled in the form of light) and heat into the surrounding material so that it can no longer cool and condense into stars and so shutting down star formation in the host galaxies once they grow large enough. The black hole activity may also be the culprit behind expelling what residual gas is left in an elliptical galaxy, so that star formation is effectively blocked after a certain point.

MAGNETIC BLACK-HOLE-FED DRAMA IN NGC 1275

This stunning image of NGC 1275 provides amazing detail and resolution of the fragile filamentary structures, which show up as a reddish lacy structure surrounding this giant elliptical galaxy. These filaments are cool despite being surrounded by gas that is at a temperature of 55 million °C. They are suspended in a magnetic field that maintains their structure and demonstrates how energy from the central black hole is transferred to the surrounding gas. This is the first time astronomers have been able to differentiate to this degree the individual 200 light-year-wide threads making up such filaments. The entire image is approximately 260,000 light-years across. Also seen in the image are impressive lanes of dust from a separate spiral galaxy that lies partly in front of the giant elliptical central cluster galaxy and has been completely disrupted by the tidal gravitational forces within the galaxy cluster. Several striking filaments of blue newborn stars are seen crossing the image.

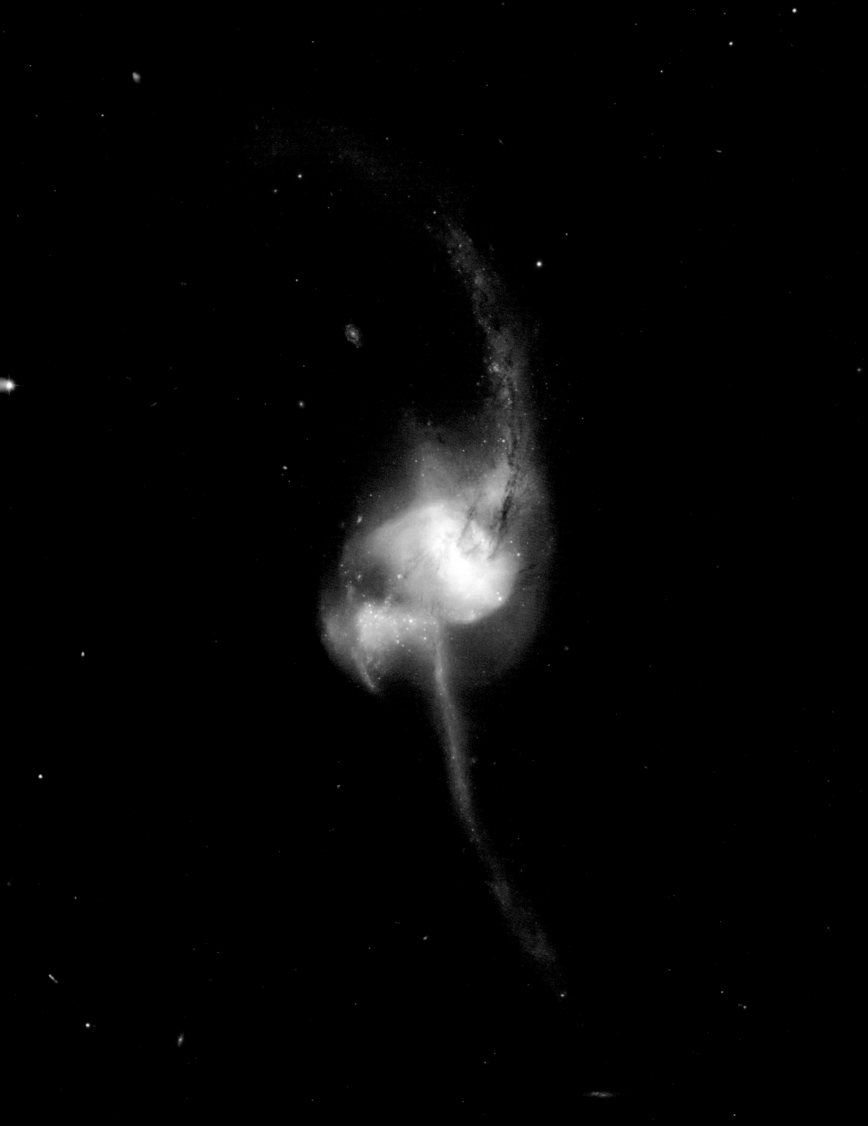

THE COLLIDING GALAXIES MOVIE

NGC 2623

Several Hubble datasets were added together to produce this deep and quite spectacular color composite of NGC 2623. The galaxy exhibits two prominent tidal tails that are 80,000 light-years long as well as a network of dust lanes. The most prominent feature in the image of NGC 2623 is a concentration of about 100 unresolved star clusters — one of the richest and most compact regions of bright star clusters known.

Having just one observing point in space, our Earth, and a very limited lifespan, we are not able to imagine fully the three-dimensional action taking place when galaxies collide, twist, and turn in space over millions of years. The sharp vision of Hubble provides us with enough snapshots of different collisions that we can piece together a movie of a full galaxy collision, frame by frame.

> *" It is clear that much of the diversity of observed galaxy collisions is due in part to the varying angles we observe them from and the different times in their lives at which we observe them."*

Imagine you could move back and forward in time and travel great distances with a speed greatly exceeding the speed of light. Observing a galaxy collision as it unfolds from start to finish over billions of years from all angles would give you a stunning view of phenomena that no one has ever seen in reality. This is naturally an impossible dream, but we can go part of the way by using supercomputers to simulate a collision using real physics. The visualization of a galaxy collision shown to the left in the panels on the following three pages is created from one single supercomputer simulation and shows the entire collision as a movie sequence from different angles. As we cannot observe the same collision over the necessary time scale we compare the simulation with *different* interacting galaxy pairs observed by Hubble. It is a bit like trying to reconstruct the cultural connections of people on Earth from just one snapshot of each city. Not an easy task, but even so the correspondence between the simulations and the pictures is stunning. This comparison of research simulations with high resolution observations allows astronomers to understand these titanic crashes better. It is clear that much of the diversity of observed galaxy collisions is due in part to the varying angles we observe them from and the different times in their lives at which we observe them.

The sequence starts with two distinct individual undisturbed spiral galaxies of roughly the same size. Gradually the two main bodies attract each other, creating shock waves that increase the process of star formation. Long tails of stars and gas are flung out by the tidal forces. Slowly the remaining structures of the galaxies are smoothed out, as the tidal tails are either lost in the surrounding space or finally rain back onto the new galaxy created in the merging. For a while longer the two supermassive black holes persist as the only testimony of the past of the combined galaxy that by now has lost its spiral patterns and structurally looks more and more like an elliptical galaxy.

Naturally the detail of a galaxy collision plays out differently depending on how evenly matched the contenders are in mass, on the approach speed, and angle compared to the spin axes of the involved galaxies. There are plenty of odd-looking systems, such as polar-ring galaxies, where the debris settles into rings over the pole of the large galaxy for a while (see page 76).

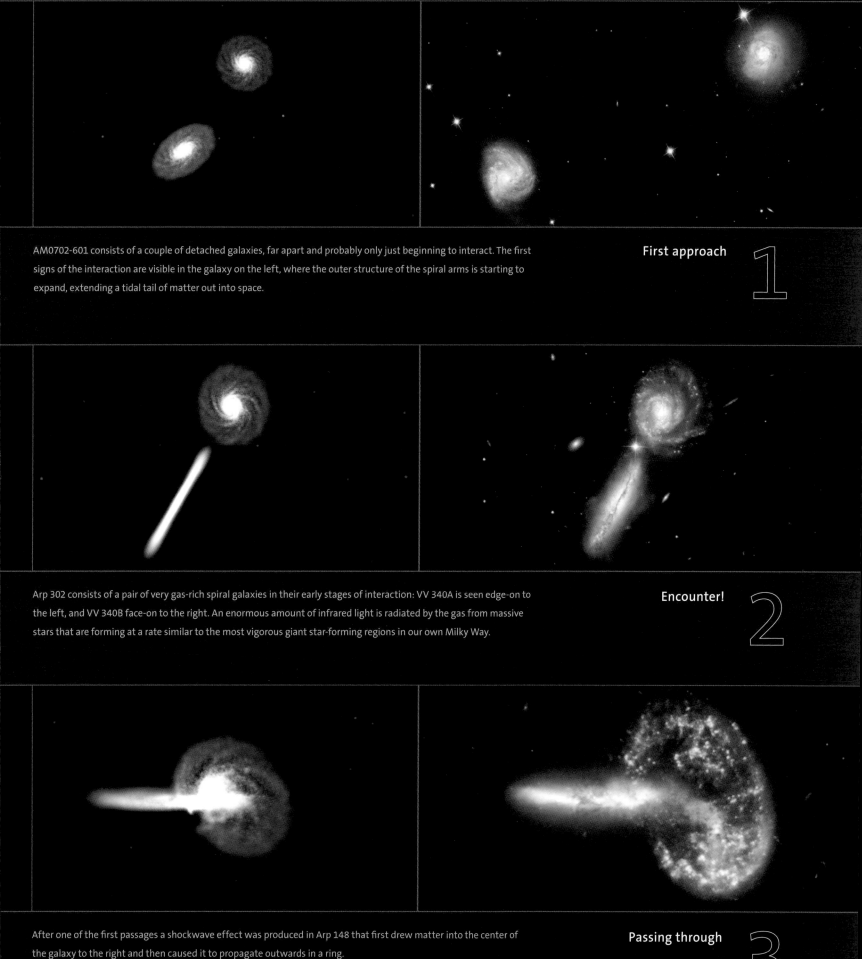

AM0702-601 consists of a couple of detached galaxies, far apart and probably only just beginning to interact. The first signs of the interaction are visible in the galaxy on the left, where the outer structure of the spiral arms is starting to expand, extending a tidal tail of matter out into space.

First approach

1

Arp 302 consists of a pair of very gas-rich spiral galaxies in their early stages of interaction: VV 340A is seen edge-on to the left, and VV 340B face-on to the right. An enormous amount of infrared light is radiated by the gas from massive stars that are forming at a rate similar to the most vigorous giant star-forming regions in our own Milky Way.

Encounter!

2

After one of the first passages a shockwave effect was produced in Arp 148 that first drew matter into the center of the galaxy to the right and then caused it to propagate outwards in a ring.

Passing through

3

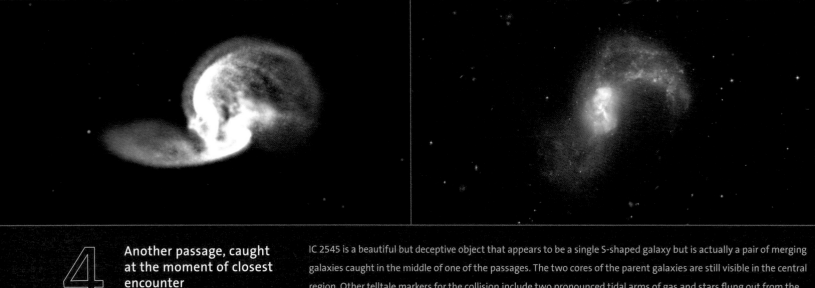

4. Another passage, caught at the moment of closest encounter

IC 2545 is a beautiful but deceptive object that appears to be a single S-shaped galaxy but is actually a pair of merging galaxies caught in the middle of one of the passages. The two cores of the parent galaxies are still visible in the central region. Other telltale markers for the collision include two pronounced tidal arms of gas and stars flung out from the central region. The tidal arm curving upwards and clockwise in the image contains a number of blue star clusters. IC 2545 lies in the constellation of Antlia, the Air Pump, some 450 million light-years from Earth.

Markarian 848 consists of two galaxies in an embracing posture, almost like a game of "Snake" where you catch your own tail. Two long, highly curved arms of gas and stars emerge from a central region with two cores. One arm, curving counter-clockwise, stretches to the bottom of the image, where it makes a U-turn and interlocks with the other arm that curves down clockwise from the top. The two cores are 16,000 light-years apart. Markarian 848 is located in the constellation of Boötes, the Bear Watcher, and is approximately 550 million light-years from Earth.

Tails interlock, cores melt together

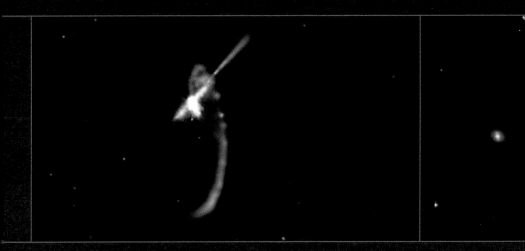

ESO 148-2 resembles an owl in flight. It consists of a pair of former disk galaxies undergoing a collision. The cores of the two individual galaxies are embedded in hot dust and contain a large number of stars. Two huge "wings" sweep out from the center and curve in opposite directions. This "cosmic owl" is one of the most luminous infrared galaxies known and is located some 600 million light-years from Earth.

Cores combined, tidal tails still raining down on the newly merged galaxy

**POLAR-RING GALAXY
NGC 4650A**

Located about 130 million light-years away, NGC 4650A is one of only 100 known polar-ring galaxies. Their unusual disk-ring structure is not yet fully understood. What remains of one of the two galaxies has become the rotating inner disk of old red stars in the center. The other, likely a smaller galaxy, had its gas stripped off and captured by the larger galaxy, forming a new ring of dust, gas, and stars that orbit around the inner galaxy almost at right angles to the old disk. This is the polar ring that we see almost edge-on.

> *"The outcome of a collision depends not only on how close the galaxies come to each other but also on how the approach angle and the spin of the galaxies align."*

The outcome of a collision depends not only on how close the galaxies come to each other but also on how the approach angle and the spin of the galaxies align. If the galaxies spin in opposite directions as they approach each other, rapid changes can occur, whereas if they rotate in the same direction they can temporarily reinforce each other's motion.

Encounters taking place near the galactic "equators" will often produce two symmetric tidal arms, while polar encounters tend to produce a very one-sided distortion. "Retrograde" encounters, when the companion moves to oppose the galaxy's sense of rotation, produce less large-scale damage and set up smaller "scalloped" ripples in a galaxy's disk.

The thin disks of spiral galaxies are what scientists call "cold" — all the stars and gas in a certain area share almost the same orbits. When disturbed, they can be pulled into very thin bridges, tails, and distorted arms, which retain their identity. In contrast, elliptical galaxies have stars moving in all directions, so their response to a gravitational disturbance yields very broad fans of stars, or an overall asymmetry. NGC 454 (page 78) shows the different responses of these two kinds of galaxies in a clear way.

Gravitational disturbances at right angles to the disk can make it warp, giving us a clue to past encounters that can remain visible even after other interaction processes take place. we see this, for example, in NGC 6670 (page 79).

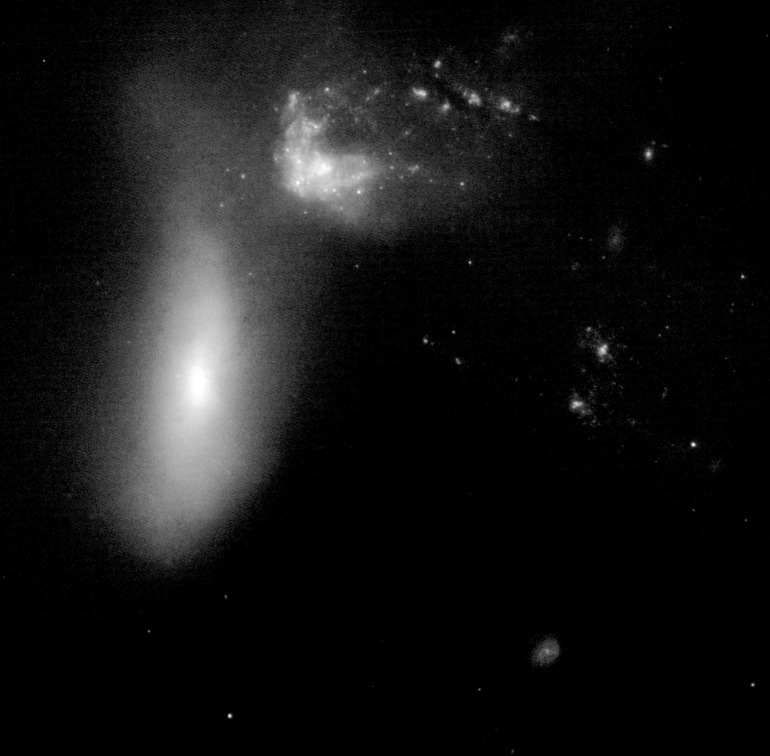

NGC 454

NGC 454 is a galaxy pair comprising a large red elliptical galaxy and an irregular gas-rich blue galaxy. The system is in the early stages of an interaction that has severely distorted both components. The three bright blue knots of very young stars to the right of the two main components are part of the irregular blue galaxy. The disturbance of the elliptical galaxy results in a broad fan of stars. The pair is approximately 164 million light-years away.

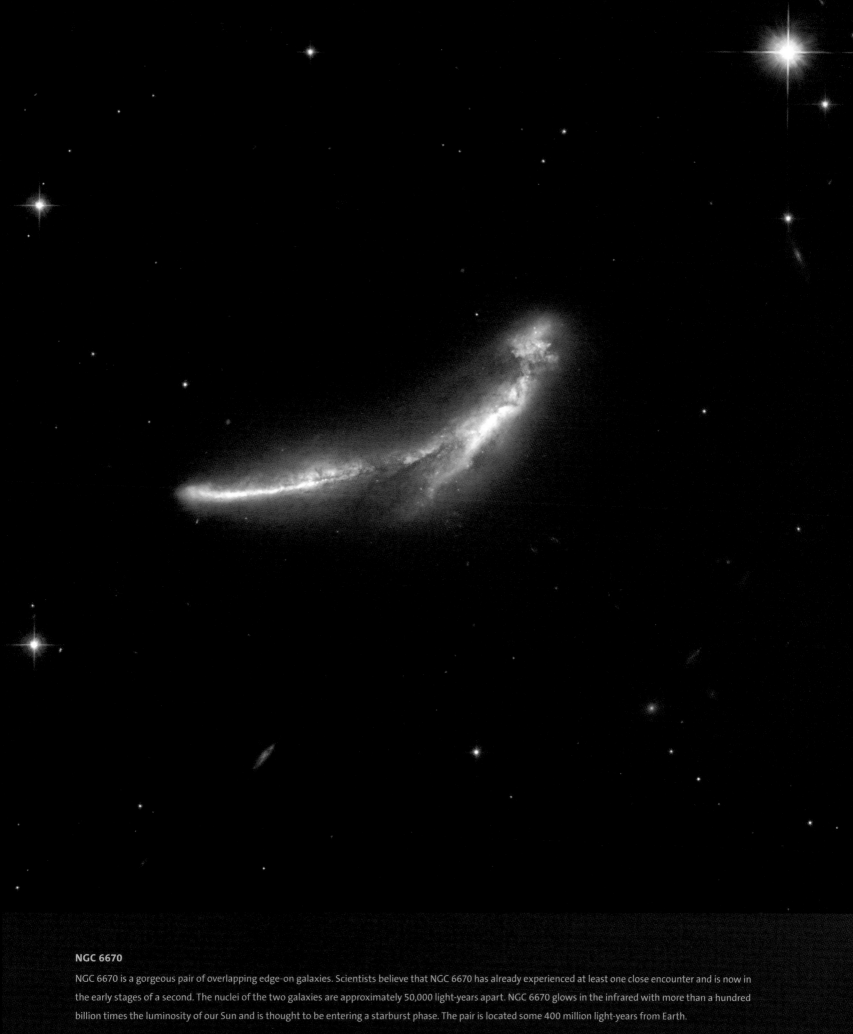

NGC 6670

NGC 6670 is a gorgeous pair of overlapping edge-on galaxies. Scientists believe that NGC 6670 has already experienced at least one close encounter and is now in the early stages of a second. The nuclei of the two galaxies are approximately 50,000 light-years apart. NGC 6670 glows in the infrared with more than a hundred billion times the luminosity of our Sun and is thought to be entering a starburst phase. The pair is located some 400 million light-years from Earth.

> *" Just as a pebble thrown into a pond creates an outwardly moving circular wave, a propagating density wave is generated at the point of impact and spreads outward. "*

Splashing Collisions

When a dense galaxy plunges through the disk of a spiral, almost perpendicular to its plane, the resulting temporary increase in the inward gravitational force creates a "splash" in the spiral's stars and gas. Just as a pebble thrown into a pond creates an outwardly moving circular wave, a propagating density wave is generated at the point of impact and spreads outward. As this density wave collides with material in the target galaxy that was moving inward due to the gravitational pull of the two galaxies, shocks and dense gas are produced, stimulating star formation. The resulting shock wave often leads to a brilliant ring of new stars, which moves outward as stars age and new ones are born. The Cartwheel Galaxy (page 82) was one of the first of these ring galaxies to be studied in detail and clever astronomers discovered that a splashing collision between two galaxies was to blame for the odd appearance of ring galaxies. Other examples of this curious phenomenon at work are seen in Arp 148 (page 81), Arp 147 (page 135), AM 0644-741 (page 102) and Hoag's Object (page 103).

Messy Mergers

Once galaxies start to merge, the complexities of their initial structures and approach make the merger itself a structural mess. Some examples from earlier are NGC 3808 (inside front cover), the infrared cousins NGC 6240 (page 63), and Arp 220 (page 62).

In the Gallery are more examples of "messy" collisions, where even the best astro-detective will be hard pressed to resolve the mystery of what happened: IC 883 (page 106), NGC 6090 (page 107), NGC 3256 (page 52), MCG+08-11-002 (page 109), NGC 5256 (page 110), UGC 5101 (page 111) and Markarian 273 (page 112).

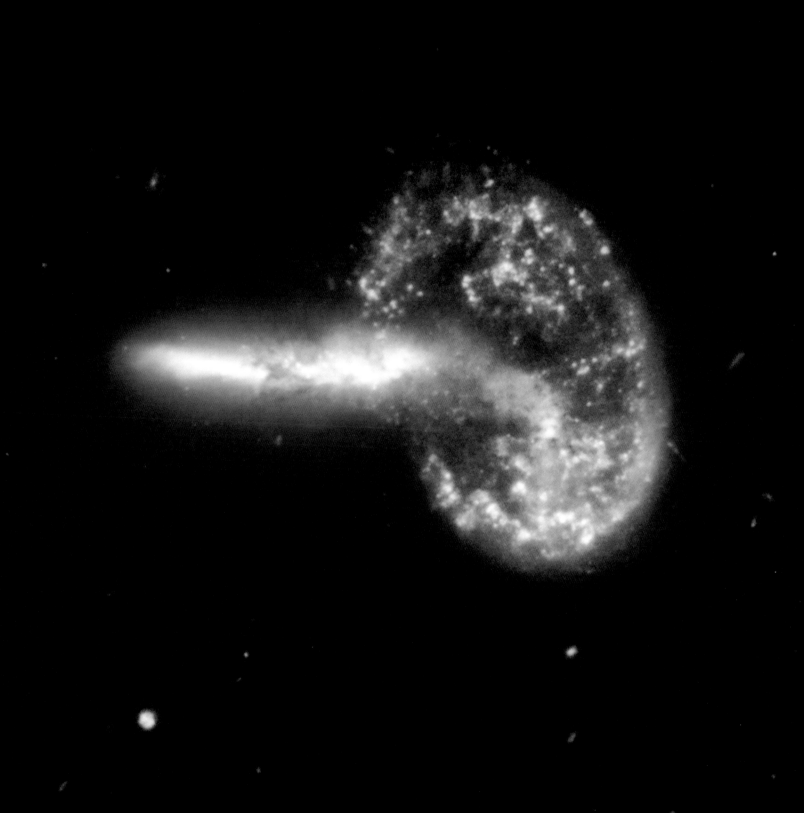

ARP 148

Arp 148 is the staggering aftermath of an encounter between two galaxies, resulting in a ring galaxy. The collision between the two parent galaxies produced a shockwave effect that first drew matter into the center and then caused it to propagate outward in a ring. Arp 148 is nicknamed "Mayall's object" and is located in the constellation of Ursa Major, the Great Bear, approximately 500 million light-years away.

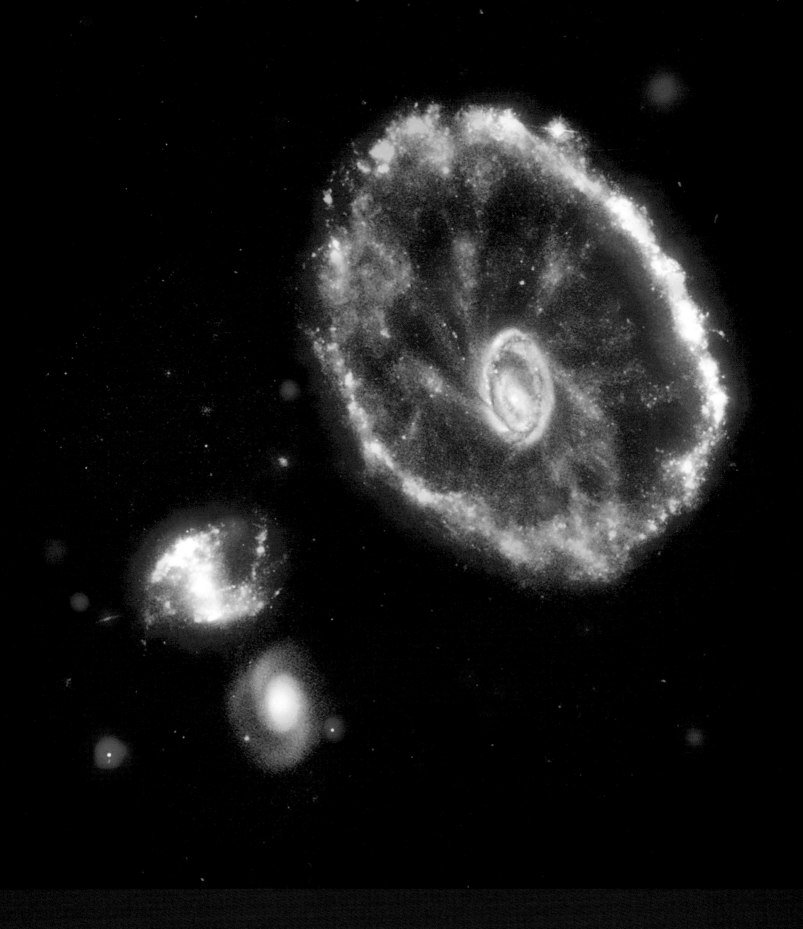

THE CARTWHEEL GALAXY

Located 500 million light-years away in the constellation Sculptor, the Cartwheel Galaxy looks like a wagon wheel. The galaxy's nucleus is the bright object in the center of the image; the spokelike structures are wisps of material connecting the nucleus to the outer ring of young stars. The galaxy's unusual configuration was created by a nearly head-on collision with a smaller galaxy about 200 million years ago. This image is a combination of exposures from the Spitzer Space Telescope (red), Hubble (green), GALEX (blue), and the Chandra X-Ray Observatory (purple).

Galaxy Groups

Interactions between galaxies are managed almost exclusively by gravity, the weakest of the four forces of nature but the only one that can act at very, very long distances. Gravity keeps us on the ground and keeps objects in the Solar System in orbit around the Sun. Gravity is also responsible for the physical relationship we have with our nearby galaxies. The Milky Way, together with the Andromeda Galaxy, the Triangulum Galaxy, and about fifty smaller galaxies, make up what is called the Local Group.

Galaxy groups with handfuls of members are not uncommon in the universe and often do — not surprisingly — show signs of disturbance.

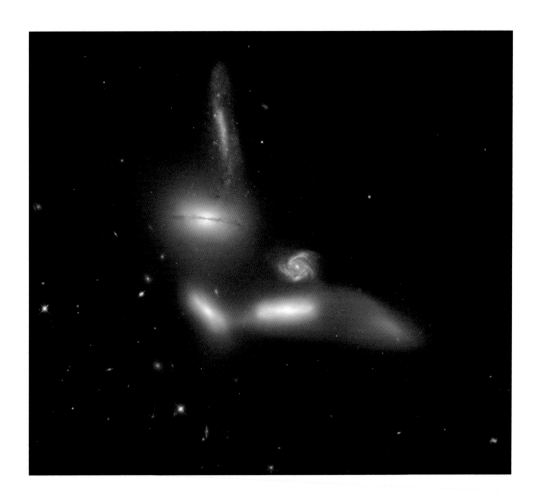

GALAXY GROUP SEYFERT'S SEXTET

Seyfert's Sextet is engaging in a slow dance of destruction that will last for billions of years. The galaxies are so tightly packed together that gravitational forces are beginning to rip stars from them and distort their shapes. These same gravitational forces eventually could bring the galaxies together to form one large galaxy. The name of this grouping, Seyfert's Sextet, implies that six galaxies are participating in the action. But only four galaxies are on the dance card. The small face-on spiral with the prominent arms in the center is a background galaxy almost five times farther away than the other four. Only a chance alignment makes it appear as if it is part of the group. The sixth member of the sextet is not a galaxy at all but a long tidal tail of stars (lower right).

HICKSON COMPACT GROUP 90

The galaxies in the Hickson Compact Group 90 are locked in a gravitational tug-of-war and are experiencing a strong tidal interaction. As a result, a significant number of stars have been ripped out of their home galaxies. These stars are now forming a luminous, diffuse component in the galaxy group.

> *" When two galaxies pass close by each other the gravitational force affecting the stars nearest to the approaching galaxy is vastly greater than that experienced by the stars at the other end. "*

Tidal Traumas

When two galaxies pass close by each other the gravitational force affecting the stars nearest to the approaching galaxy is vastly greater than that experienced by the stars at the other end. This difference causes the stars to be pulled in opposite directions and is called the tidal effect. On Earth a similar effect rules the twice-a-day ebb and flow of the oceans (and the crust). In this case, the effect is induced by the difference of the gravitation "felt" on Earth at its nearest and furthest distances to the Moon and similarly, but to a lesser degree, to the Sun.

Tides between galaxies are, however, much more disruptive than ocean tides because the stars are more loosely bound to the galaxy than the components of Earth, which are bound together by chemical forces. The tidal forces between two colliding galaxies can reshape and rip the galaxies apart in such an encounter. The resulting tidal traumas cause bridges, antennas, arcs, tails, and more. Stars and gas are flung into space.

The longer the interaction lasts, the bigger and more dramatic are the disruptions. So, the consequences of a cosmic encounter are often more severe if the relative velocity of the two interacting galaxies is low — the opposite of what is normal on Earth.

Since the gravitational force is directly proportional to the involved masses, it could seem that it is only the mass that rules the hierarchy in the close encounters between stellar systems. This is, however, not always true. In the interactions between galaxies the compactness also plays a role. So, given two galaxies of roughly the same masses but very different sizes, the larger one, where the matter is less densely packed, will probably suffer the most serious consequences.

ARP 240

Arp 240 is an astonishing galaxy pair, composed of spiral galaxies of similar mass and size, NGC 5257 and NGC 5258. The galaxies are visibly interacting with each other via a tidal bridge of dim stars connecting the two galaxies, almost like two dancers holding hands while performing a pirouette. Both galaxies harbor supermassive black holes in their centers and are actively forming new stars in their disks. Arp 240 is located in the constellation Virgo, approximately 300 million light-years away.

THE MICE

A spectacular pair of galaxies engaged in a celestial dance of cat and mouse or, in this case, mouse and mouse. Located 300 million light-years away in the constellation Coma Berenices, the colliding galaxies have been nicknamed "The Mice" because of the long tidal tails of stars and gas emanating from each galaxy. Otherwise known as NGC 4676, the pair will eventually merge into a single giant galaxy.

Destroying Dwarfs

Edwin Hubble first coined the term "Local Group" in his book, *The Realm of the Nebulae*, to describe those galaxies that were isolated in the general field but were in the vicinity of the Milky Way. With the increased sensitivity of modern telescopes and detectors the Local Group of galaxies has grown, with many new smaller members identified. The search has mostly been carried out by sensitive large-area surveys of halos of the Milky Way and the Andromeda Galaxy that have looked for slight over-densities of stars. This has made it possible to locate dozens of very faint dwarf satellites that have previously eluded detection.

Some of these dwarf galaxies are clearly marked by interactions with the other galaxies in the Local Group. As small intruders they can be shredded by the tidal forces unleashed by an approaching massive galaxy, leaving behind a trail of debris. In our own galaxy dozens of stellar streams have been found. These are former dwarf galaxies that have been extruded into long, fuzzy, spaghetti-like strands of stars that wrap around the Milky Way. During each pass of the Milky Way more stars are stripped away from the passing intruder and join the halo of our galaxy. It is thought that these kinds of events may be quite common in the evolution of most large galaxies.

NGC 5907 WITH DEBRIS OF A DWARF GALAXY

In this ground-based amateur image astronomers have for the first time obtained visual proof of ongoing galactic cannibalism. These debris structures surround the galaxy NGC 5907, located 40 million light-years from Earth and formed from the destruction of one of its dwarf satellite galaxies. The dwarf galaxy has lost the greater part of its mass in the form of stars, star clusters, and dark matter, all of which are strewn out all along its orbit, giving rise to a complicated assembly of criss-crossing galactic fossils whose radius exceeds 150,000 light-years.

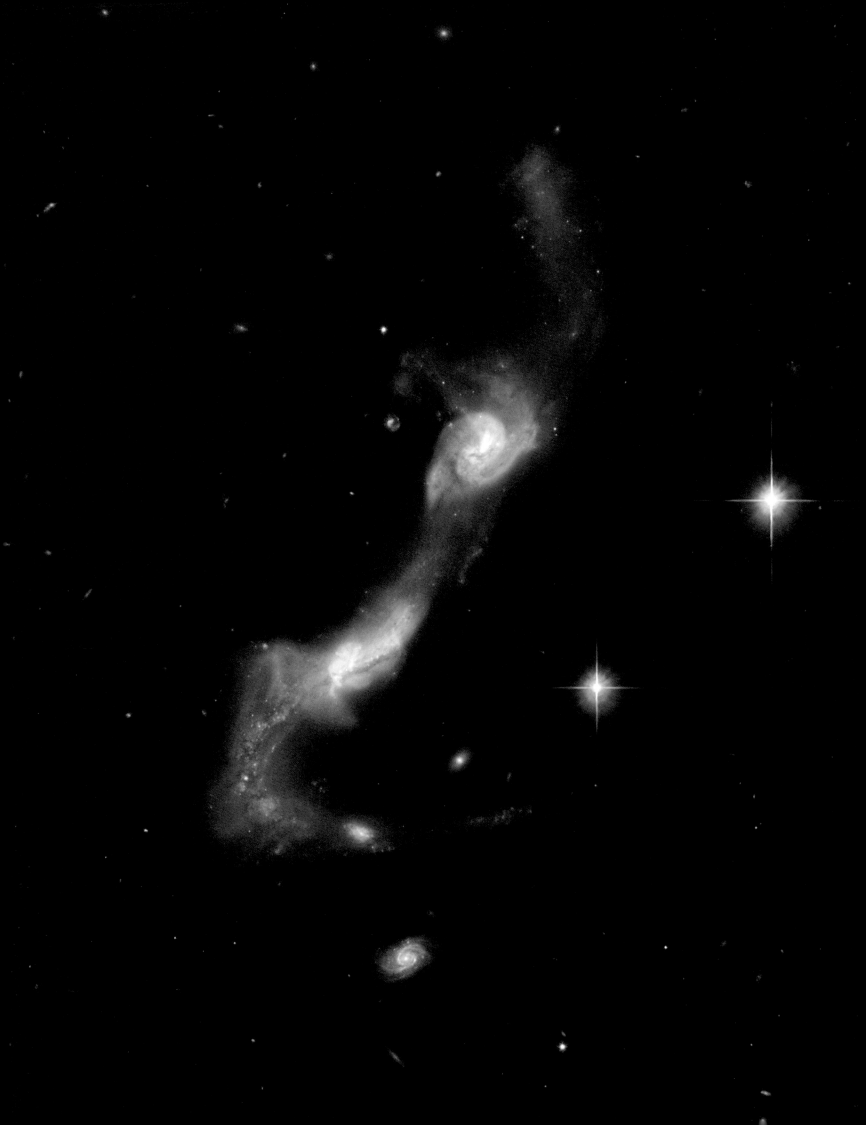

5 | THE END

UGC 8335
UGC 8335 is a strongly interacting pair of spiral galaxies resembling two ice skaters. The interaction has united the galaxies via a bridge of material and has yanked two strongly curved tails of gas and stars from the outer parts of their "bodies." Both galaxies have dust lanes in their centers. UGC 8335 is located in the constellation of Ursa Major, the Great Bear, about 400 million light-years from Earth.

A fuzzy cloud of light in the constellation of Andromeda connects us to our future. The fate of the entire Milky Way Galaxy, and perhaps Earth itself, is intertwined with the Andromeda Galaxy. The latest astronomical evidence strongly suggests that the two giant spiral members of our Local Group are on a collision course, with a merger likely occurring in five billion years.

Galaxies were once thought of as "island universes," evolving slowly in complete isolation. This is now known not to be the case. Galaxies interact in a variety of ways with satellite and neighboring galaxies, and collisions and mergers of galaxies are now believed to be the key evolutionary mechanism. There are probably very few galaxies in the universe that were not shaped by interactions, mergers, and acquisitions. The position of a galaxy in the Hubble sequence likely depends strongly on the number and severity of collisions in its past history. Spirals, formed from relatively isolated primordial gas clouds, appear at one end of the sequence, and giant ellipticals produced through the mergers of spirals and smaller galaxies, appear at the other. In between, mergers between galaxies of differing mass produce a variety of galaxies that can take all kinds of forms as they are merging.

Milkomeda, the End of the Milky Way

The Andromeda Galaxy is one of the most awe-inspiring astronomical objects visible in the northern sky. Though probably known since antiquity, it was first noted as a nebulous patch in the tenth century by the Persian astronomer Abd al-Sufi. We now know that this faint patch of light is actually a great spiral galaxy, an island of hundreds of billions of stars located about 2.5 million light-years away. It is the most distant object humans can see with the naked eye. Remarkably, the light we see from there began its journey at the dawn of humanity.

It is expected that, in the far future, our Milky Way will collide with the Andromeda Galaxy, our nearest large neighbor, generating a spectacle unmatched in our corner of the universe. Milkomeda is the name some astronomers have given to the end product of this merger.

The two galaxies are approaching each other at almost 500,000 kilometers per hour and, in three to five billion years, will collide head-on. The direct collision will lead to a magnificent merger between the two galaxies, during which the Milky Way will no longer be the spiral galaxy we are familiar with. Instead, together with the Andromeda Galaxy, over the course of a billion years it will evolve into a single, huge, elliptical galaxy.

As the galaxies approach each other, strong tidal forces will seriously deform the stately spiral structures, sending stars careening in all directions as their galactic orbits are perturbed. Even though these galaxies contain hundreds of billions of stars, individual stars will not collide because they are very small relative to the separation between them. But interstellar gas clouds will collide and collapse, triggering furious episodes of starbirth in both galaxies.

New calculations made in 2008 by Thomas Cox and Abraham Loeb (Harvard-Smithsonian Center for Astrophysics) give a good idea of the fate of the Sun and Solar System in the process.

Computer modeling shows that the two galaxies will make their first close pass only two billion years from now. It will likely take another two to three billion years before they merge. The five billion years that may elapse before the collision is over is a timescale comparable to the lifetime of the Sun, which may give future astronomers in the Solar System the chance to witness parts of the spectacle. But it is unfortunately likely that, on its way to becoming a red giant, the Sun will have already grown large enough to boil away the Earth's oceans and fry all life by the time the merger is complete. If not, any surviving descendants of humans will have a very different view from today. The Milky Way band in the night sky will be replaced by the huge elliptical-shaped glow of billions of stars.

During the collision, the two galaxies will swing around each other a couple of times, intermingling their stars as gravitational forces mix them together. There is a chance (or risk if you will) that the Sun will be ejected into a long tidal tail following the first passage of the Andromeda Galaxy. It is also altogether possible that during parts of the violent encounter the Sun will be more tightly bound to Andromeda. Future astronomers in the Solar System might then see the Milky Way as an external galaxy in the night sky. Other effects of the merger would likely be enhanced comet showers due to the many stars passing by the Solar System and nudging the comet-filled Oort Cloud, as well as a firework of newborn stars in the night sky, possibly paving the way for new planets with new civilizations capable of building space telescopes to look for their origin.

THE ANDROMEDA GALAXY

Due to the Andromeda Galaxy's large size on the night sky Hubble can only take pictures of a small portion of it. This beautiful ground-based photo was taken by hobby astrophotographer Robert Gendler. Most of the individual stars seen are foreground stars in our Milky Way. Two of the Andromeda galaxy's companion galaxies, Messier 32 and Messier 110 are also seen.

MILKOMEDA
Artist's impression of the spectacular encounter of the Andromeda Galaxy and the Milky Way, which will take place in the next few billion years. Milkomeda is the nickname astronomers have given to the endproduct of this merger.

6 GALLERY

NGC 4921

The Coma Galaxy Cluster is one of the closest very rich collections of galaxies in the nearby universe. The galaxies in rich clusters undergo many interactions and mergers that tend to gradually turn gas-rich spirals into elliptical systems without much active star formation. NGC 4921 is one of the rare spirals in the Coma Cluster and a rather unusual one — with just a delicate swirl of dust in a ring around the galaxy. Much of the pale spiral structure in the outer parts of the galaxy is unusually smooth and gives the whole galaxy the ghostly look of a vast translucent jelly-fish. An extraordinary rich background of more remote galaxies stretching back to the early universe is also seen.

Interacting galaxies are found throughout the universe, sometimes as dramatic collisions that trigger bursts of star formation, on other occasions as stealthy mergers that result in new galaxies. Each of the various merging galaxies in this gallery is a snapshot of a different instant in the long interaction process. Many of the Hubble images seen here were taken as part of a large investigation of luminous and ultraluminous infrared galaxies called the Great Observatories All-sky LIRG Survey.

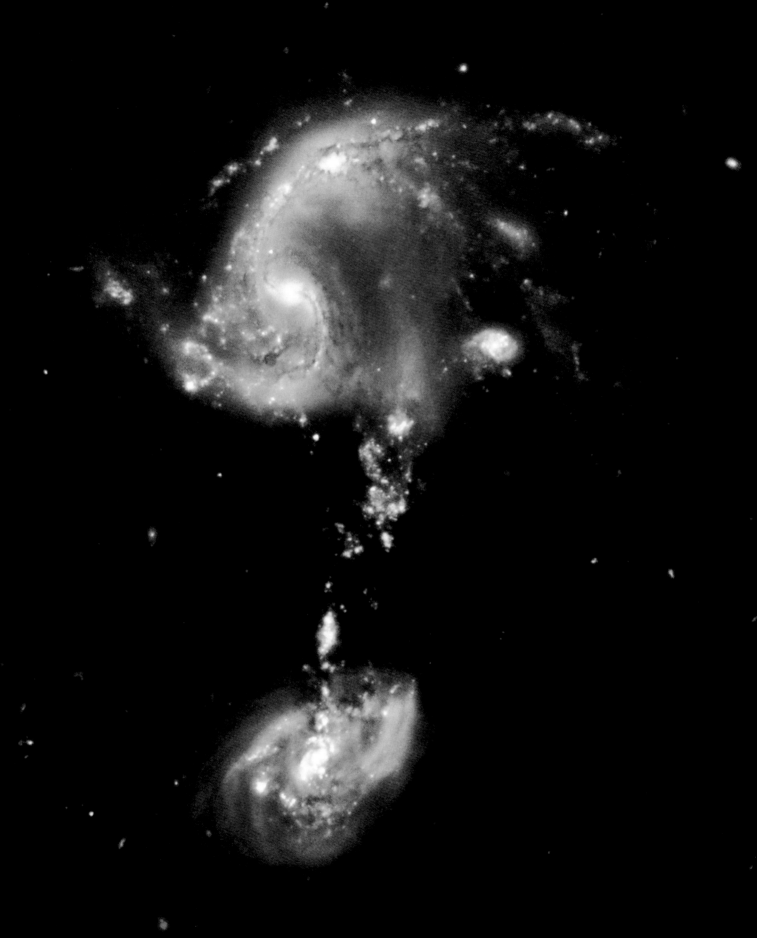

ARP 194

This uniquely interacting group contains two distinct components connected by a cosmic fountain of stars, gas, and dust that stretches over 100,000 light years. The compact bright starbirth regions are young super star clusters that formed as a result of the interaction. The gravitational forces involved in a galaxy interaction can enhance the star formation rate and give rise to luminous bursts of star formation in merging systems.

ARP 274

Arp 274 is a pair of close-knit galaxies that look like they are shaking hands — or rather spiral arms. Under the relentless pull of gravity, the galaxies weave elegant, twisted lanes of dust and stars, as well as brilliant blue clusters of newborn stars and pinkish stellar nurseries. A pair of foreground stars inside our own Milky Way is seen near the bottom.

RING GALAXY AM 0644-741

Resembling a diamond-encrusted bracelet, a ring of brilliant blue star clusters wraps around the yellowish nucleus of what was once a normal spiral galaxy. The sparkling blue ring is 150,000 light-years in diameter, making it larger than the entire Milky Way. This lies 300 million light-years away in the direction of the southern constellation of Volans.

HOAG'S OBJECT

A nearly perfect ring of hot blue stars pinwheels about the yellow nucleus of an unusual galaxy known as Hoag's Object. The entire galaxy is about 120,000 light-years across, which is slightly larger than our Milky Way Galaxy. The blue ring, which is dominated by clusters of young, massive stars, contrasts sharply with the yellow nucleus of mostly older stars, and is likely made up of the shredded remains of one of the two galaxies that collided. Curiously, an object that bears an uncanny resemblance to Hoag's Object can be seen in the gap at the one o'clock position. The object is probably a background ring galaxy.

ESO 69-6

The galaxies of this beautiful interacting pair bear some resemblance to musical notes. Long tails sweep out from the two galaxies, caused by gas and stars that were stripped out and sheared away from the outer regions of the galaxies. These are the unique signatures of an interaction. ESO 69-6 is located in the constellation of Triangulum Australe, the Southern Triangle, about 650 million light-years away from Earth.

ESO 99-4

ESO 99-4 is a galaxy with a highly peculiar shape that is probably the remnant of an earlier merger process that deformed it beyond visual recognition, leaving the main body largely obscured by dark bands of dust. ESO 99-4 lies in a rich field of foreground stars, in the constellation of Triangulum Australe, the Southern Triangle, about 400 million light-years away.

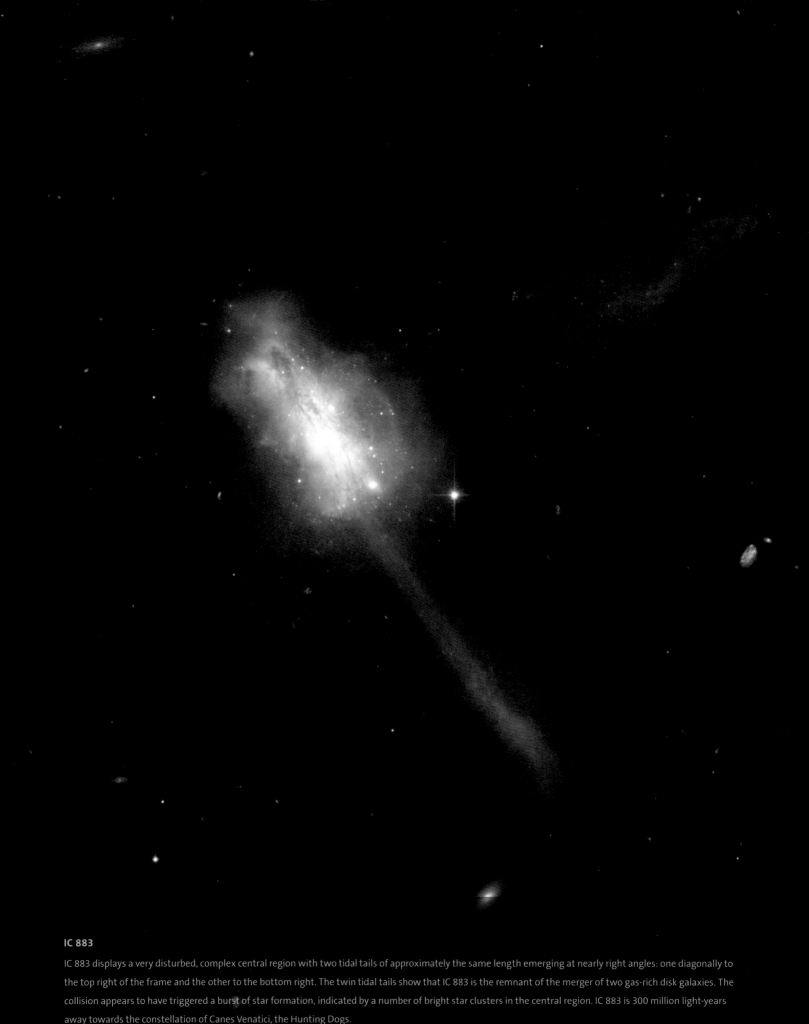

IC 883

IC 883 displays a very disturbed, complex central region with two tidal tails of approximately the same length emerging at nearly right angles: one diagonally to the top right of the frame and the other to the bottom right. The twin tidal tails show that IC 883 is the remnant of the merger of two gas-rich disk galaxies. The collision appears to have triggered a burst of star formation, indicated by a number of bright star clusters in the central region. IC 883 is 300 million light-years away towards the constellation of Canes Venatici, the Hunting Dogs.

NGC 6090

NGC 6090 is a beautiful pair of spiral galaxies with an overlapping central region and two long tidal tails formed from material ripped out of the galaxies by gravitational interaction. The two visible cores are approximately 10,000 light-years apart, suggesting that the two galaxies are at an intermediate stage in the merging process. The Hubble image reveals bright knots of newborn stars in the region where the two galaxies overlap. The upper component has a clear spiral structure and is viewed face-on, while the other, just below, is seen edge-on with no spiral arms visible. NGC 6090 is located in the constellation of Draco, the Dragon, about 400 million light-years from Earth, and shares many features with the famous Antennae galaxies (see the cover).

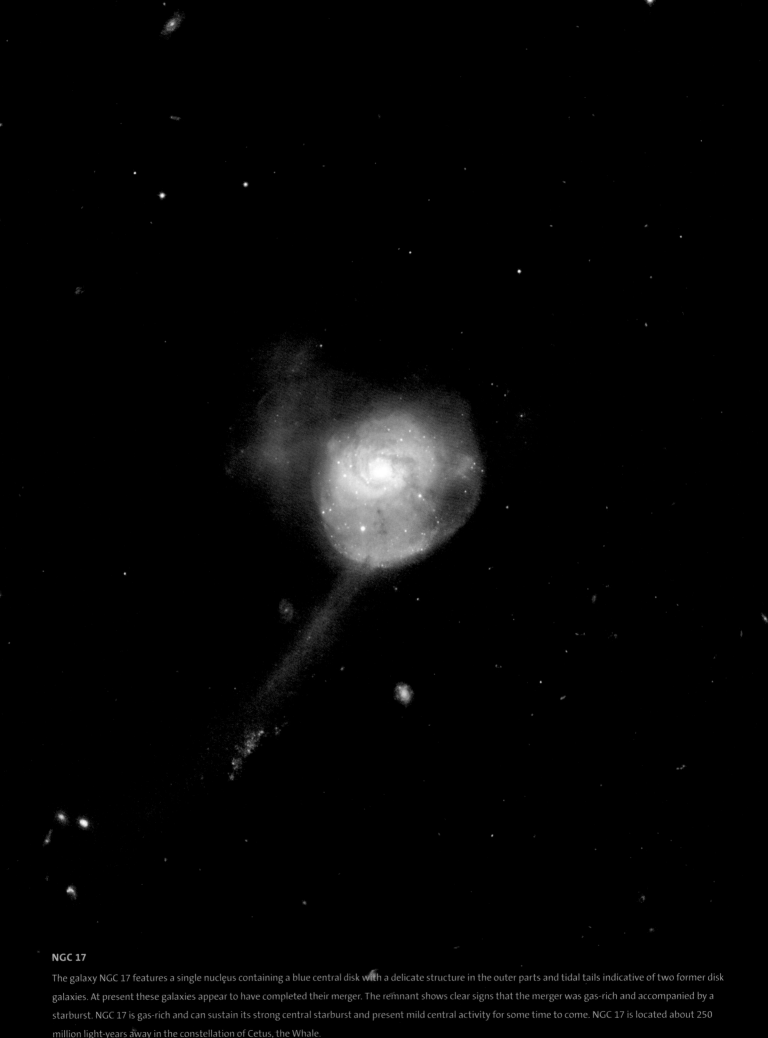

NGC 17

The galaxy NGC 17 features a single nucleus containing a blue central disk with a delicate structure in the outer parts and tidal tails indicative of two former disk galaxies. At present these galaxies appear to have completed their merger. The remnant shows clear signs that the merger was gas-rich and accompanied by a starburst. NGC 17 is gas-rich and can sustain its strong central starburst and present mild central activity for some time to come. NGC 17 is located about 250 million light-years away in the constellation of Cetus, the Whale.

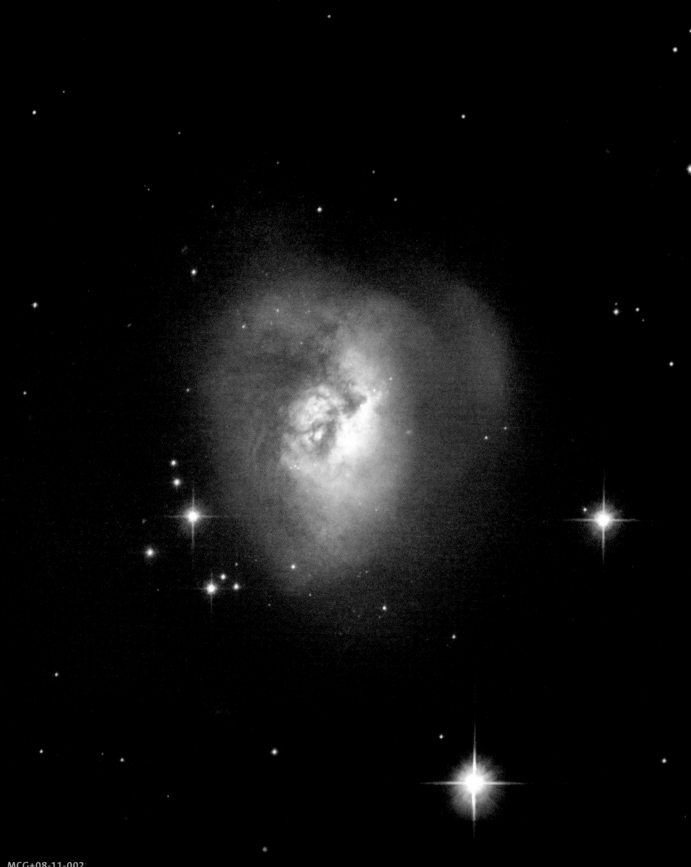

MCG+08-11-002

MCG+08-11-002 is an odd-looking galaxy with a spectacular dark band of absorbing dust in front of the galaxy's center, making it resemble a "Black Eye." This peculiar galaxy is at the center of a rich field of foreground stars, close to the plane of our own Milky Way Galaxy. MCG+08-11-002 is about 250 million light-years away in the constellation of Auriga, the Charioteer.

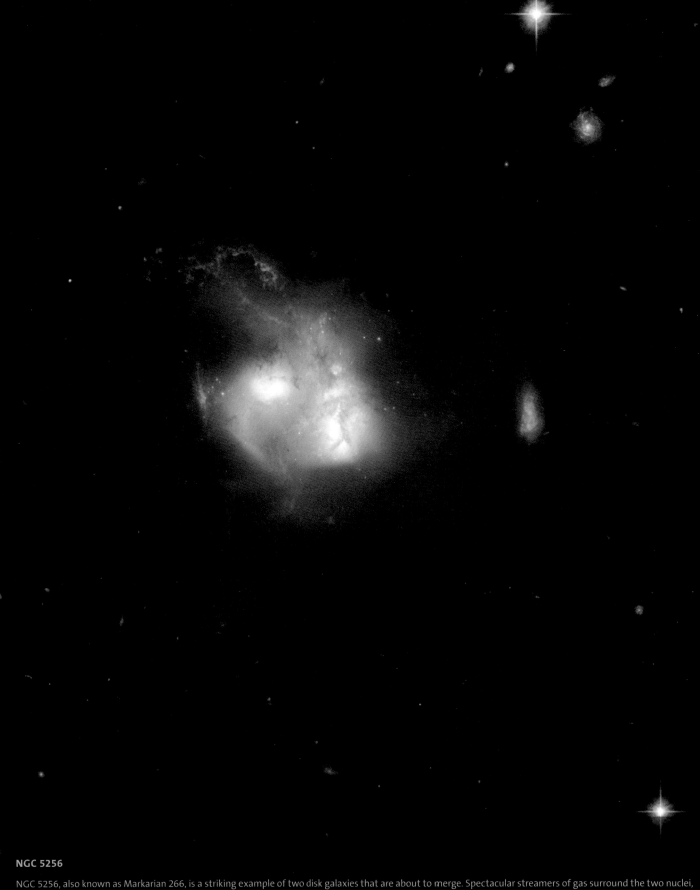

NGC 5256

NGC 5256, also known as Markarian 266, is a striking example of two disk galaxies that are about to merge. Spectacular streamers of gas surround the two nuclei, and eye-catching blue spiral trails indicate recent star formation. The shape of the object is highly disturbed, and observations in various wavelength regimes — infrared, millimeter-wave and radio — provide additional evidence for a starburst in this system. NGC 5256 is located in the constellation of Ursa Major, the Great Bear, some 350 million light-years from Earth. Each galaxy also contains an active galactic nucleus, evidence that the chaos is allowing gas to fall into the regions around central black holes as well as feeding starbursts. Recent observations from the Chandra X-ray Observatory show that both nuclei, as well as a region of hot gas between them, have been heated by the shock waves produced as gas clouds at high velocities collide.

UGC 5101

UGC 5101 is a peculiar galaxy with a single nucleus contained within an unstructured main body that suggests a recent interaction and merger. NGC 5101 is thought to contain an active galactic nucleus. A pronounced tidal tail extends diagonally to the top-left of the frame. A fainter halo of stars surrounds the galaxy and is visible in the image, due to Hubble's ability to collect and detect faint light. This halo is probably a result of the earlier collision. UCG 5101 is about 550 million light-years from Earth.

MARKARIAN 273

Markarian 273 is a galaxy with a bizarre structure that somewhat resembles a toothbrush. The Hubble image shows an intricate central region and a striking tidal tail that extends diagonally towards the bottom-right of the image. The tail is about 130,000 light-years long and is strongly indicative of a merger between two galaxies. Markarian 273 has an intense region of starburst, where 60 solar masses of new stars are born each year. Near-infrared observations reveal a nucleus with two components. Markarian 273 is one of the most luminous galaxies when observed in the infrared, and is located 500 million light-years from Earth.

ESO 593-8

ESO 593-8 is an impressive pair of interacting galaxies with a feather-like galaxy crossing a companion galaxy. The two components will probably merge to form a single galaxy in the future. The pair is adorned with a number of bright blue star clusters. ESO 593-8 is located in the constellation of Sagittarius, the Archer, some 650 million light-years from Earth.

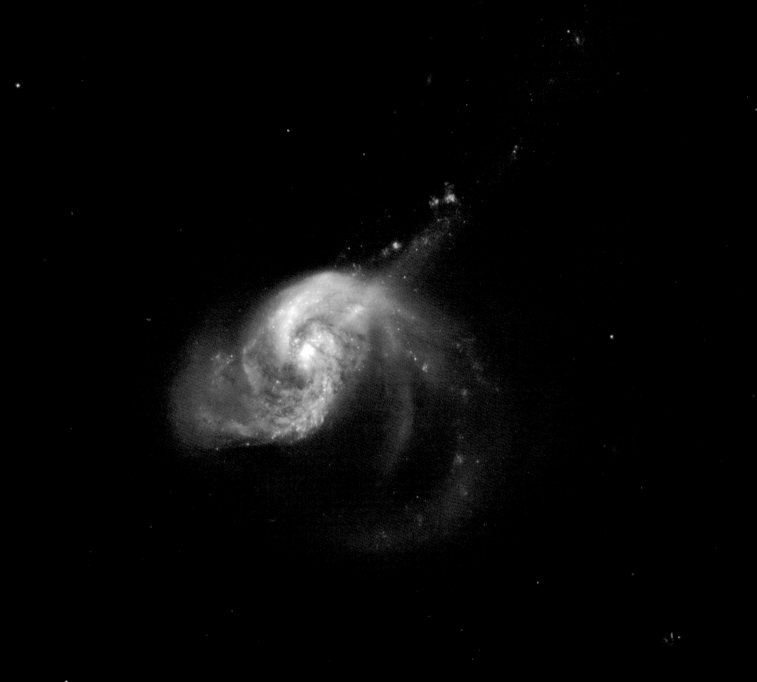

NGC 1614

The galaxy system NGC 1614 has a bright optical center and two clear inner spiral arms that are fairly symmetrical. It also has a spectacular outer structure that consists principally of a large, one-sided, curved extension of one of these arms to the lower right, and a long, almost straight tidal tail that emerges from the nucleus and crosses the extended arm to the upper right.

MARKARIAN 231

The extraordinary galaxy Markarian 231 was discovered in 1969 as part of a survey searching for galaxies with strong ultraviolet radiation. It has long tidal tails and a disturbed shape. Results from the first spectrum showed clear signs of the presence of a powerful quasar in the center that made Markarian 231 unique in the Markarian sample. Markarian 231 has maintained its reputation as an exceptional object since those early observations and continues to be a favorite target in all wavelength regimes. Its infrared luminosity is similar to that of quasars, making it one of the most luminous and powerful of the known ultraluminous infrared galaxies. Although the emission of many ultraluminous infrared galaxies appears to be dominated by energetic starbursts, Markarian 231 has been repeatedly identified as an exception, and many pieces of evidence point toward an accreting black hole as the major power source behind the enormous infrared luminosity.

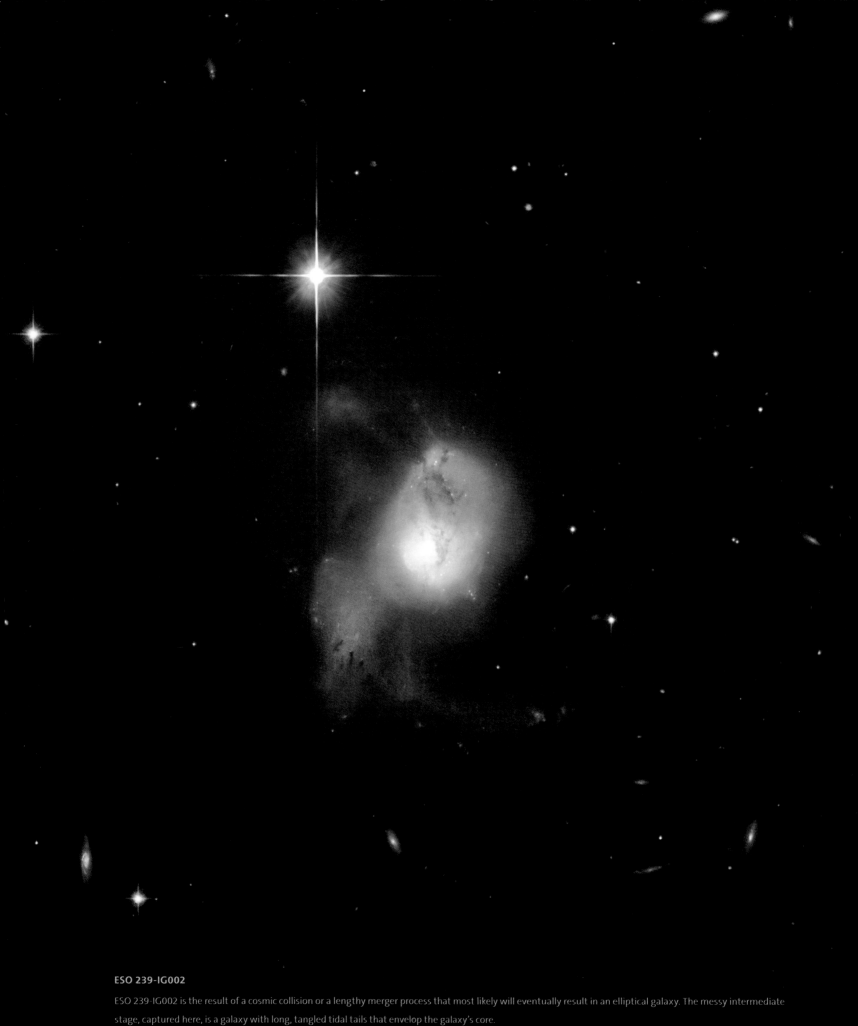

ESO 239-IG002

ESO 239-IG002 is the result of a cosmic collision or a lengthy merger process that most likely will eventually result in an elliptical galaxy. The messy intermediate stage, captured here, is a galaxy with long, tangled tidal tails that envelop the galaxy's core.

ZWICKY II 96

Zwicky II 96 is a system of merging galaxies with a bizarre shape. Powerful young starburst regions appear as long threadlike structures hanging between the main galaxy cores. The system almost qualifies as an ultraluminous system but has not yet reached the late stage of coalescence that is the norm for most ultraluminous systems. Zwicky II 96 is located in the constellation of Delphinus, the Dolphin, about 500 million light-years from Earth.

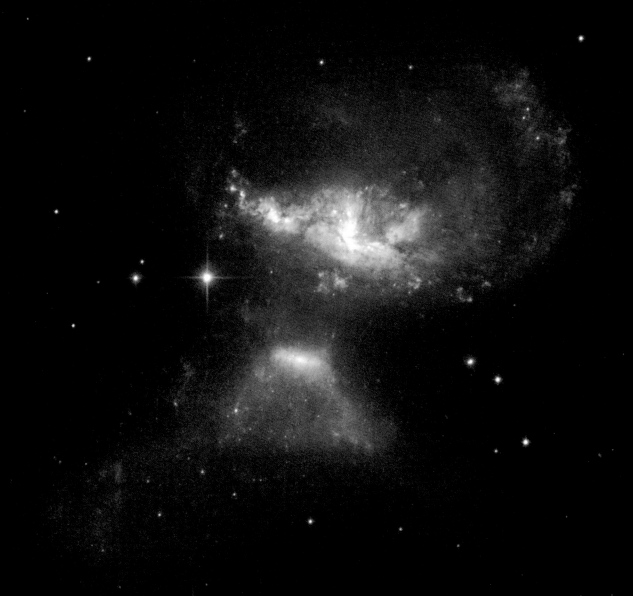

MCG+12-02-001

MCG+12-02-001 consists of a pair of galaxies visibly affected by gravitational interaction as material is flung out in opposite directions. A large galaxy can be seen at the top of the frame and a smaller galaxy resembling an erupting volcano is at the bottom. The bright core of this galaxy emerges from the summit of the "volcano." MCG+12-02-001 is a luminous infrared system that radiates with more than 100 billion times the luminosity of our Sun. It is located some 200 million light-years from Earth toward the constellation of Cassiopeia, the Seated Queen.

MARKARIAN 533

NGC 7674 (seen just above the center), also known as Markarian 533, is the brightest and largest member of the so-called Hickson 96 compact group of galaxies, consisting of four galaxies. This stunning Hubble image shows a spiral galaxy nearly face-on. The central bar-shaped structure is made up of stars. The shape of NGC 7674, including the long narrow streamers seen to the left of and below the galaxy, can be accounted for by tidal interactions with its companions. NGC 7674 has a powerful active nucleus that is perhaps fed by gas drawn into the center through interactions with the companions. It is located in the constellation of Pegasus, the Winged Horse, about 400 million light-years from Earth.

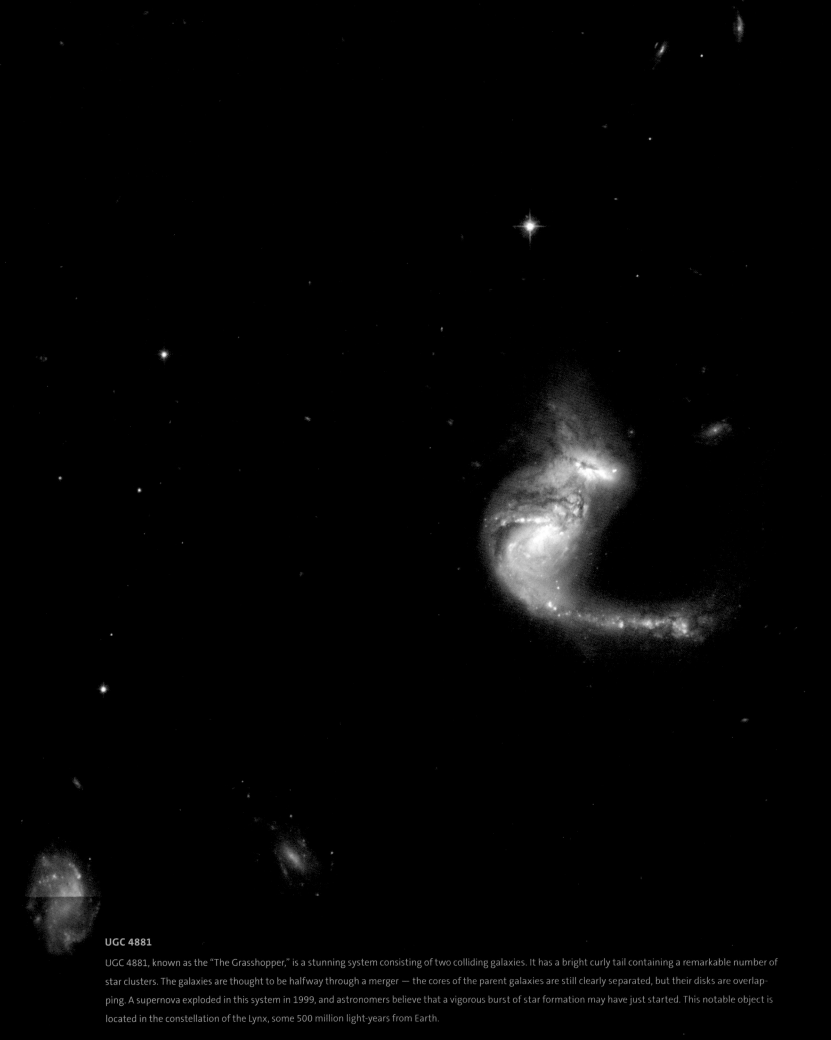

UGC 4881

UGC 4881, known as the "The Grasshopper," is a stunning system consisting of two colliding galaxies. It has a bright curly tail containing a remarkable number of star clusters. The galaxies are thought to be halfway through a merger — the cores of the parent galaxies are still clearly separated, but their disks are overlapping. A supernova exploded in this system in 1999, and astronomers believe that a vigorous burst of star formation may have just started. This notable object is located in the constellation of the Lynx, some 500 million light-years from Earth.

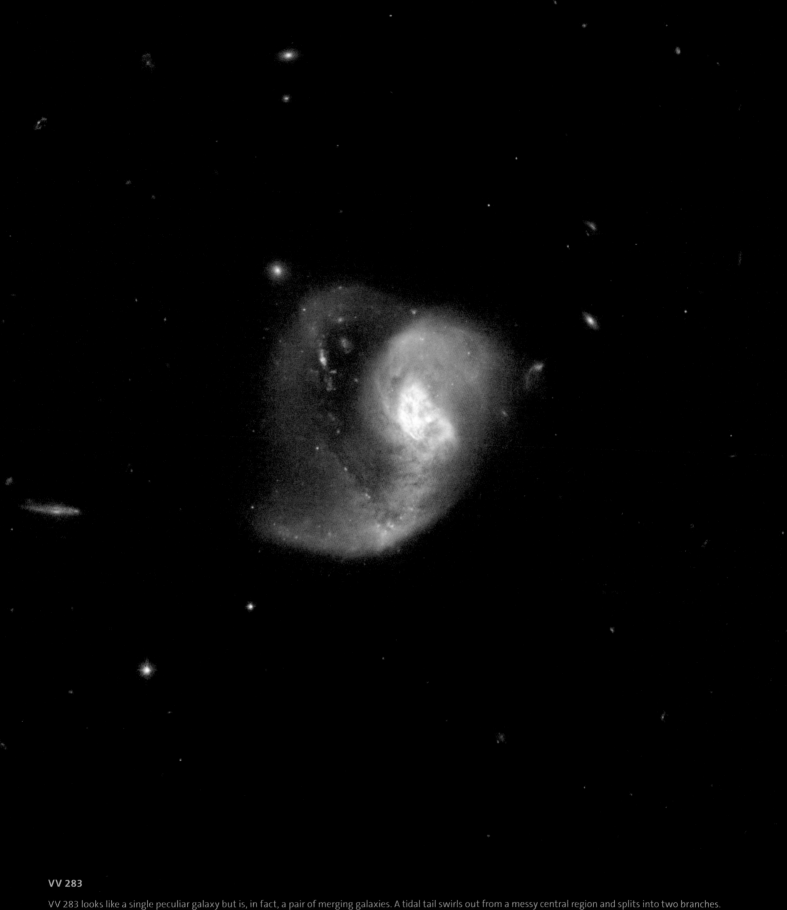

VV 283

VV 283 looks like a single peculiar galaxy but is, in fact, a pair of merging galaxies. A tidal tail swirls out from a messy central region and splits into two branches. The upward twisting branch is brightened by luminous blue star knots. Like many merging systems, VV 283 is a highly luminous infrared system, radiating nearly 1,000 billion times energy more than our Sun. VV 283 is located in the constellation of Virgo, the Maiden, some 500 million light-years away.

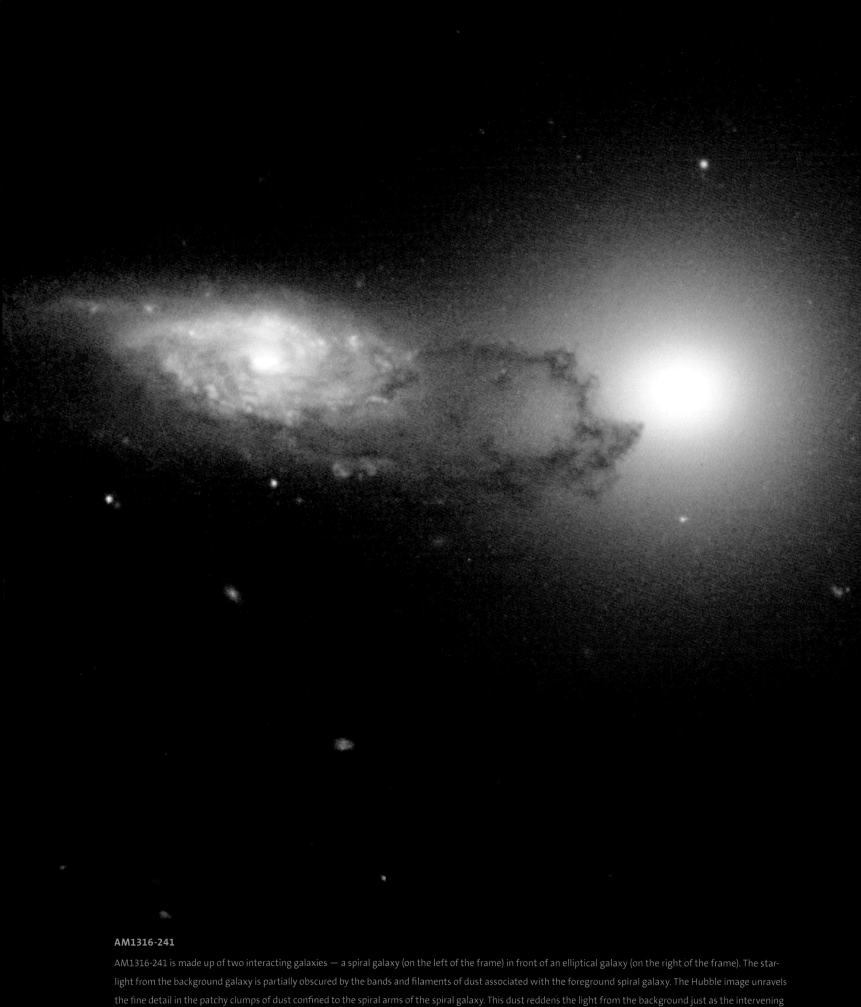

AM1316-241

AM1316-241 is made up of two interacting galaxies — a spiral galaxy (on the left of the frame) in front of an elliptical galaxy (on the right of the frame). The starlight from the background galaxy is partially obscured by the bands and filaments of dust associated with the foreground spiral galaxy. The Hubble image unravels the fine detail in the patchy clumps of dust confined to the spiral arms of the spiral galaxy. This dust reddens the light from the background just as the intervening dust in Earth's atmosphere reddens sunsets here. AM1316-241 is located some 400 million light years away toward the constellation of Hydra, the Water Snake.

IC 4687

IC 4687 forms a triplet with two other galaxies: IC 4686 to the right and IC 4689 further to the right. IC 4687 has a chaotic body of stars, gas, and dust and a large curly tail to the left. The two companions are partially obscured by dark bands of dust. The interacting triplet is about 250 million light-years from Earth, in the constellation of Pavo, the Peacock.

IRAS 20351+2521

IRAS 20351+2521 is a galaxy with a sprawling structure of gas, dust, and numerous blue star knots. It is located in the constellation of Vulpecula, the Fox, 450 million light-years from Earth.

NGC 7469

This is a stunning pair of interacting galaxies, the barred spiral galaxy NGC 7469 (Arp 298, Markarian 1514), a luminous infrared source with a powerful starburst deeply embedded in its circumnuclear region, and its smaller companion IC 5283. This system is located about 200 million light-years from Earth in the constellation of Pegasus, the Winged Horse.

NGC 5754 EXHIBITING ITS SPIRAL ARMS

This beautiful pair of interacting galaxies consists of NGC 5754, the large spiral on the top and NGC 5752, the smaller companion in the bottom left corner of the image. NGC 5754's inner structure has hardly been disturbed by the interaction, but the outer structure does exhibit tidal features, as does the symmetry of the inner spiral pattern and the kinked arms just beyond its inner ring. In contrast, NGC 5752 has undergone a starburst episode, with a rich population of massive and luminous star clusters clumped around the core and intertwined with intricate dust lanes. The contrasting reactions of the two galaxies to their interaction are due to their differing masses and sizes. NGC 5754 is located in the constellation of Boötes, the Herdsman, some 200 million light-years away.

AM0500-620

AM0500-620 consists of a highly symmetric spiral galaxy seen nearly face-on and partially backlit by a background galaxy. The foreground spiral galaxy has a number of dust lanes between its arms. The background galaxy was earlier classified as an elliptical galaxy, but Hubble has now revealed it to have dusty spiral arms and bright knots of stars. AM0500-620 is 350 million light-years from Earth in the constellation of Dorado, the Swordfish.

ESO 550-IG02

ESO 550-IG02 shows a pair of spiral galaxies, the larger nearly face-on and accompanied by a smaller, highly tilted partner. Tidal interaction from the smaller companion has clearly deformed one arm of the larger galaxy. Strong star formation continues both in the deformed arm and in a ring structure around the galaxy's core. The pair is surrounded by the glow of faintly shining stars and interstellar matter that has been smeared through space by the gravitational effects of the collision and the pull of a third, nearby galaxy.

LEDA 62867 AND NGC 6786

This image displays a beautiful pair of interacting spiral galaxies with swirling arms. The smaller of the two, dubbed LEDA 62867 and positioned to the left of the frame, seems to be safe for now, but will probably be swallowed by the larger spiral galaxy, NGC 6786 (to the right), eventually. There is already some disturbance visible in both components. A supernova was seen to explode in the large spiral in 2004. NGC 6786 is located in the constellation of Draco, the Dragon, about 350 million light-years away.

ESO 507-70

ESO 507-70 is an odd-looking remnant of an earlier merger process. It is a chaotic swirl of gas, dust, and stars, with no sign of the original spiral or elliptical structure, and it appears lost and distorted beyond recognition after the gravitational encounter with other galaxy. ESO 507-70 is some 300 million light-years from Earth toward the constellation of Hydra, the Water Snake.

NGC 5331

NGC 5331 is a pair of interacting galaxies beginning to "link arms." There is a blue trail in the image that flows to the right of the system. NGC 5331 is very bright in the infrared, with about 100 billion times the luminosity of the Sun. It is located in the constellation of Virgo, the Maiden, about 450 million light-years from Earth.

2MASXJ09133888-1019196

2MASXJ09133888-1019196 comprises two interacting galaxies that are both disturbed by gravitational interaction. The wide separation of the pair — approximately 130,000 light-years — suggests that the galaxies are just beginning to merge. Together the two galaxies form an ultraluminous infrared system, which is unusual for the early stages of an interaction. One possible explanation is that one or both of the components have already experienced a merger or interaction. Giant black holes lurk at the cores of both galaxies, which are found in the constellation of Hydra, the Sea Serpent, about 700 million light-years from Earth.

IRAS 21101+5810

This system is an interacting galaxy pair. The interaction has disturbed both galaxies. The lower galaxy has a bizarre structure, and a tidal tail emerges from the main body of the upper galaxy. The galaxy pair lies in a crowded field of Milky Way stars. IRAS 21101+5810 is located in the constellation of Cepheus, the King, about 550 million light-years from Earth.

ARP 272

Arp 272 is a remarkable collision between two spiral galaxies, NGC 6050 and IC 1179, and is part of the Hercules Galaxy Cluster, located in the constellation of Hercules. The galaxy cluster is part of the Great Wall of clusters and superclusters, the largest known structure in the universe. The two spiral galaxies are linked by their swirling arms. Arp 272 is located some 450 million light-years away from Earth.

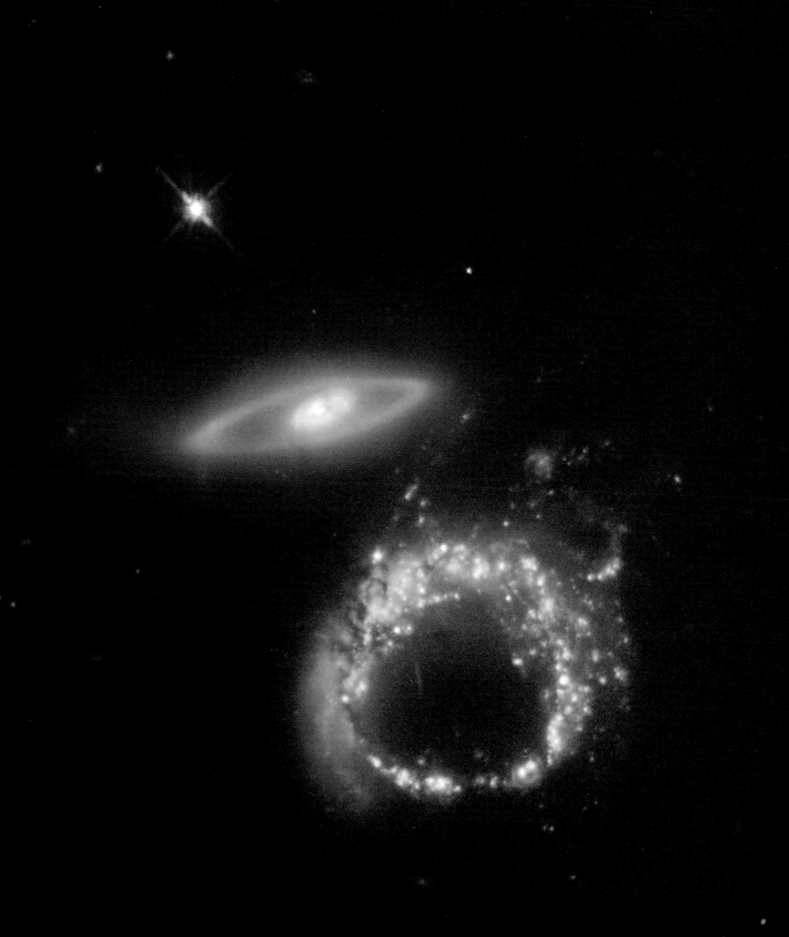

ARP 147

Arp 147 is a particularly intriguing pair of gravitationally interacting galaxies. The blue ring of intense star formation was most probably formed after the galaxy on the left passed through the galaxy on the right. The dusty reddish knot at the lower left of the blue ring probably marks the location of the original nucleus of the galaxy that was hit.

THE AUTHORS

Lars Lindberg Christensen

Lars is a science communication specialist heading the European Southern Observatory education and Public Outreach Department in Munich, Germany, where he is responsible for public outreach and education for the VLT, La Silla, for ESO's part of ALMA (the largest and most expensive ground-based astronomical project currently under construction), E-ELT (the largest planned visible light/near-infrared telescope) and ESA's part of the Hubble Space Telescope.

He obtained his Master's Degree in physics and astronomy from the University of Copenhagen, Denmark and has more than 100 publications to his credit, most of them in popular science communication and its theory. His other productive interests cover several major areas of communication, including graphical, written, technical and scientific communication. He has written six books, including *Eyes on the Skies, Hidden Universe, The Hands-On Guide for Science Communicators* and *Hubble — 15 Years of Discovery*. His books have been translated to Finnish, Portuguese, Danish, German, Korean and Chinese.

Lars is Press Officer for the International Astronomical Union (IAU), a founding member and secretary of the IAU Commission 55 Communicating Astronomy with the Public, manager of the ESA/ESO/NASA Photoshop FITS Liberator project, executive editor of the peer-reviewed Communicating Astronomy with the Public journal, director of the Hubblecast and ESOcast video podcast, manager of the IAU International Year of Astronomy 2009 Secretariat and the Executive producer and director of the science documentaries *Eyes on the Skies* and *Hubble — 15 Years of Discovery*. In 2005 Lars was the youngest recipient so far of the Tycho Brahe Medal for his achievements in science communication. Lars lives in Garching near Munich, Germany, with his wife and son.

Raquel Yumi Shida

Raquel started on the road to the cosmos as an amateur astronomer in Brazil, her home country, having made her own telescope in her teenage years and providing observational data and images of various astronomical objects to scientists and organizations worldwide.

Her first participation in Hubble projects was back when she was an undergraduate student, when she was awarded a research internship at the Space Telescope Science Institute in the USA. Before joining the team of ESA's Hubble group in Germany in 2006, she worked at the Astronomy Department at the University of São Paulo, Brazil, in the development of pioneer and award-winning educational projects involving the use of a network of automated telescopes. At the same university, she graduated with a professional degree in architecture and urban planning.

Currently, she is part of a group of scientists, astronomy communicators, and designers at the European Southern Observatory education and Public Outreach Department in Germany. She focuses on the use of internet-based technologies for bringing a wide variety of visual products and important astronomical news from Hubble, ESO, the International Astronomical Union and the International Year of Astronomy 2009 to the general public.

In her spare time, she still enjoys making amateur astronomy observations and photography.

Davide De Martin

Davide is an electrical engineer working in Venice for the largest Italian power supply company. He is responsible for the company's fleet vehicle management and is involved in the development and implementation of new technologies and methods to improve work-force management.

Davide has been an amateur astronomer since his childhood, when he started to devour everything about astronomy. He has been an editor of the Italian magazine *Coelum* since 1996, writing dozens of columns and articles ranging from the history of astronautics to astronomy popularization. He has realized or collaborated in a variety of astronomy popularization projects as websites, multimedia supports, shows, exhibits and books.

Recently, Davide has mainly focused his activities on processing old data from observatory archives and developing techniques to turn them into images. Most of his work is presented at www.skyfactory.org.

He has been part of the European Southern Observatory education and Public Outreach Department as an image processing team member since 2005, producing some of the most popular Hubble and ESO images. Although Davide has become used to looking at the universe through Hubble's optics, he still loves the direct eye contact with the night-sky using his Dobsonian telescope.

RESOURCES

Arp, H. 1966, *The Atlas of Peculiar Galaxies*, NASA's Extragalactic Database (Caltech): http://ned.ipac.caltech.edu/level5/Arp/frames.html

Arp, H., Madore B. F., Roberton, W. 1987: *A Catalogue of Southern Peculiar Galaxies and Associations (Cambridge University Press)*: http://ned.ipac.caltech.edu/level5/SPGA_Atlas/frames.html

Barnes, J.E. & Hernquist, L. 1992, *Dynamics of Interacting Galaxies*, Annu. Rev. Astron. Astrophys., 30, 705,42

Carroll, B. W. & Ostlie, D. A. 2007, *An Introduction to Modern Astrophysics*, Chapter 26: Galactic Evolution, (Addison Wesley)

Christensen, L.L. & Fosbury, R. A. E. 2005, *Hubble — 15 Years of Discovery*, (New York: Springer)

Christensen, L. L., Fosbury, R. A. E. & Hurt, R. 2008, *Hidden Universe*, (Wiley)

Dubinsky, J. 2006, *The Great Milky Way-Andromeda Collision*, Sky & Telescope, October 2006

Günther, V. T. 2007, *Kosmische Unfälle*, Sterne und Weltraum, May 2007

Holmberg, E., 1941, *On the Clustering Tendencies among the Nebulae. II. a Study of Encounters Between Laboratory Models of Stellar Systems by a New Integration Procedure.* Astrophysical Journal (http://esoads.eso.org/abs/1941ApJ....94..385H)

Jayawardhana, R. 2008, *How the Milky Way devours its neighbors*, Astronomy, March 2008

Kanipe, J. & Webb, D. 2006, *The Arp Atlas of Peculiar Galaxies* (Willmann-Bell)

Keel, William: personal web page: http://www.astr.ua.edu/keel/

Leverington, D. 1996, *A History of Astronomy — from 1890 to the Present*, (New York: Springer)

Schilling, G & Christensen, L. L. 2008, *Eyes on the Skies — 400 Years of Telescopic Discovery*, (Wiley)

Toomre, A. & Toomre, J., 1972, *Galactic Bridges and Tails*, Astrophysical Journal 178, 623, (http://esoads.eso.org/abs/1972ApJ...178..623T)

Vorontsov-Velyaminov, B.A 1959, *Atlas and Catalogue of interacting galaxies*, Part 1, http://www.sai.msu.su/sn/vv/ and http://ned.ipac.caltech.edu/level5/VV_Cat/frames.html

Wikipedia articles:

http://en.wikipedia.org/wiki/Galaxy

http://en.wikipedia.org/wiki/Galaxy_formation_and_evolution

http://en.wikipedia.org/wiki/Atlas_of_Peculiar_Galaxies

IMAGE CREDITS

Cover
NASA, ESA, and the Hubble Heritage Team STScI/AURA-ESA/Hubble Collaboration. Acknowledgement: B. Whitmore (Space Telescope Science Institute) and James Long (ESA/Hubble).

Inside Front Cover
A. Fujii, NASA, ESA, Digitized Sky Survey 2 and The Hubble Heritage Team (STScI/AURA)

p. 4, 5, 52, 62, 66, 67, 78 , 86, 90, 104, 105, 106, 108, 109, 110, 111, 112, 113, 114, 115, 116, 117, 118, 119, 120, 121, 123, 124, 125, 126, 128, 129, 130, 131, 132, 133
NASA, ESA, the Hubble Heritage Team (STScI/AURA)-ESA/Hubble Collaboration and A. Evans (University of Virginia, Charlottesville/NRAO/Stony Brook University)

p. 6
NASA

p. 8, 21, 32, 35, 39, 65, 76, 102, 103
NASA, ESA, and the Hubble Heritage Team (STScI/AURA)

p. 11
Jacopo Tintoretto/The Yorck Project

p. 12-13
ESO/S. Brunier

p. 14
History of Science Collections, University of Oklahoma Libraries; copyright the Board of Regents of the University of Oklahoma

p. 16
NASA/JPL-Caltech/R. Hurt

p. 17, 34, 44, 70, 83
European Space Agency & NASA

p. 23
NASA, ESA and the Hubble Heritage (STScI/AURA)-ESA/Hubble Collaboration

p. 24, 28, 29
NASA/ESA

p. 26
NASA/ESA, Jeffrey Kenney (Yale University), Elizabeth Yale (Yale University)

p. 30
NASA, ESA, and A. Zezas (Harvard-Smithsonian Center for Astrophysics); GALEX data: NASA, JPL-Caltech, GALEX Team, J. Huchra et al. (Harvard-Smithsonian Center for Astrophysics); Spitzer data: NASA/JPL-Caltech/Harvard-Smithsonian Center for Astrophysics.

p. 37
NASA, ESA, A. Aloisi (ESA/STScI) and The Hubble Heritage (STScI/AURA)-ESA/Hubble Collaboration

p. 38
NASA, ESA, and A. Aloisi (European Space Agency and Space Telescope Science Institute)

p. 40
NASA, ESA, and S. Beckwith (STScI) and the HUDF Team

p. 42
NASA, ESA, and Johan Richard (Caltech, USA). Acknowledgement: Davide De Martin & James Long (ESA/Hubble)

p. 47
X-ray: NASA/CXC/CfA/M.Markevitch et al.; Optical: NASA/STScI; Magellan/U.Arizona/D.Clowe et al.; Lensing Map: NASA/STScI; ESO WFI; Magellan/U.Arizona/D.Clowe et al.

p. 49
NASA, ESA, G. Miley and R. Overzier (Leiden Observatory) and the ACS Science Team

p. 51
NASA, ESA and the Hubble Heritage (STScI/AURA)-ESA/Hubble Collaboration. Acknowledgment: M. West (ESO, Chile)

p. 55
NASA, ESA, the Hubble Heritage Team (STScI/AURA)-ESA/Hubble Collaboration and W. Keel (University of Alabama, Tuscaloosa)

p. 56
William Parsons/NASA, ESA, S. Beckwith (STScI), and The Hubble Heritage Team STScI/AURA)

p. 58
Reproduced by permission of the AAS

p. 59
NASA, ESA, the Hubble Heritage Team (STScI/AURA)-ESA/Hubble Collaboration and B. Whitmore (STScI)

p. 60
Reproduced by permission of the AAS

p. 63
NASA, ESA, JPL-Caltech

p. 68
NASA, ESA, and G. Canalizo (University of California, Riverside)

p. 69
NASA, ESA and Andy Fabian (University of Cambridge, UK)

p. 73, 74, 75
Frank Summers/STScI & NASA, ESA, the Hubble Heritage Team (STScI/AURA)-ESA/Hubble Collaboration and A. Evans (University of Virginia, Charlottesville/NRAO/Stony Brook University)

p. 79
NASA, ESA, the Hubble Heritage Team (STScI/AURA)-ESA/Hubble Collaboration and A. Evans (University of Virginia, Charlottesville/NRAO/Stony Brook University)

p. 81
NASA, ESA, the Hubble Heritage Team (STScI/AURA)-ESA/Hubble Collaboration and A. Evans (University of Virginia, Charlottesville/NRAO/Stony Brook University), K. Noll (STScI), and J. Westphal (Caltech)

p. 82
NASA/ESA/JPL-Caltech/P. N. Appleton (SSC/Caltech)

p. 84
NASA/ESA, J. English (U. Manitoba), S. Hunsberger, S. Zonak, J. Charlton, S. Gallagher (PSU), and L. Frattare (STScI)

p. 87
NASA, Holland Ford (JHU), the ACS Science Team and ESA

p. 89
R. Jay Gabany

p. 94-95
Robert Gendler

p. 96-97
Sky & Telescope illustration by Casey Reed

p. 99
NASA, ESA and K. Cook (Lawrence Livermore National Laboratory, USA)

p. 100, 101
The NASA/ESA Hubble Space Telescope

p. 107
NASA, ESA, the Hubble Heritage Team (STScI/AURA)-ESA/Hubble Collaboration and A. Evans (University of Virginia, Charlottesville/NRAO/Stony Brook University) and G. Ostlin (Stockholm University)

p. 122, 127
NASA, ESA, the Hubble Heritage Team (STScI/AURA)-ESA/Hubble Collaboration and W. Keel (University of Alabama, Tuscaloosa)

p. 134
NASA, ESA, the Hubble Heritage Team (STScI/AURA)-ESA/Hubble Collaboration and K. Noll (STScI)

p. 135
NASA, ESA, and M. Livio (STScI)

Inside Back Cover
NASA, Holland Ford (JHU), the ACS Science Team and ESA

Printed in the United States of America

3995

**Freeport Memorial Library
Freeport, N.Y. 11520
516-379-3274**